BLADESMITHING FOR BEGINNERS

Make Your First Knife in 7 Steps

TABLE OF CONTENTS

INTRODUCTION

One of the earliest knives made from smelted metal was discovered in a tomb in Anatolia. It was dated to the year 2,500 BC. However, iron was not included as a vital material in the making of knives until the year 500 BC.

Since then, the metal took new importance in crafting of various tools and implements. Greeks, Celts, Egyptians, and Vikings began to use iron into their metalworks. It was not until the development and discovery of steel that metalwork took on a whole new form.

We have come a long way since the first time man discovered the use of iron. Back then, the tools and knives that were produced were crude implements. They were there to serve a necessity and made using simple techniques.

Today, knife making is a process. It starts with finding the right steel, forging the knife, then subjecting the tool to an annealing and normalizing process. It is then shaped with grinding, heat treated, quenched, and finally tempered.

While our ancestors may not have been particularly careful about their working conditions, we should always make sure that we protect ourselves.

- Implement the use of safety glasses to protect our eyes from unwanted materials - such as hot metal and sharp debris - from flying into our eyes.

- Hearing protection is vital since prolonged exposure to loud sounds (for example, the noise of metal grinding) can affect hearing.

- Use a respirator to protect from tiny dust and other particles that can enter the lungs and cause permanent damage.

- Do not wear shorts, even on hot days. Hot sparks can fly off the metal and burn the skin.

- Put on leather bibs so that any stray spark hits a layer of fireproof material rather than your clothes.

- Tie up long hair when working with tools and metals. Make sure to secure a longer beard or keep them away using other means during metalworking.

- Since you are just starting, get comfortable using gloves. Eventually, they can become optional as you gain experience working with metals. But for now, better to err on the side of caution. Note: Do not use gloves while using any sort of spinning tool like a buffer or grinder. They could get caught in the mechanism.

Most importantly, have fun in the process. Don't be afraid to experiment. After all, it is only through experimentation that you can find out what you should do and what you are not supposed to do. Familiarize yourself with the basics and put the things you learn into practice.

Mistakes happen. That's alright. You may find out that your knives break, or you have an oddly shaped knife in your hands. Remember, there are only two things that are going to happen during knife making:

- You make a knife, whether it is perfect or not.
- You learn a lesson.

The lessons you learn are one of the most important aspects of knife making. You go from making regular knives to creating something like the one below.

Figure 1: Knife with tasteful curves.

So do not easily despair when you are working on your knife. Keep your focus on what you want to achieve, and you will get there eventually.

Take your time to understand the steps and information provided in this book. Make sure you know what you are working with and keep trying until you perfect it.

Design, Stock removal, grinding bevels, heat treatment, adding the handle, polishing, sharpening – these are the 7 steps with which you will make your first knife. But it's not the what, but the how to that will make you learn this craft.

With that, let's dive deeper into the art of bladesmithing.

FREE BONUSES FOR THE READERS

First of all, I want to congratulate you on taking the right steps to learn and improve your bladesmithing skills, by buying this book.

Few people take action on improving their craft, and you are one of them.

This book has exhaustive knowledge on bladesmithing and will help you make your first knife.

However, to get the most out of this book, I have 3 resources for you that will REALLY kickstart your knife making process and improve the quality of your knives.

Since you are now a reader of my books, I want to extend a hand, and improve our author-reader relationship, by offering you all 3 of these bonuses for FREE.

All you have to do is visit **https://www.elitebladesmithingmasterclass.com** and enter the e-mail where you want to receive these resources.

These bonuses will help you:

1. Make more money when selling your knives to customers
2. Save time while knife making

Here's what you receive for FREE:

1. Bladesmith's Guide to Selling Knives
2. Hunting Knife Template
3. Stock Removal Cheat Sheet

Here is a brief description of what you will receive in your inbox:

1. Bladesmith's Guide to Selling Knives

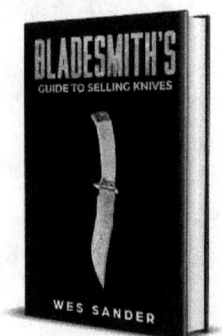

Do you want to sell your knives to support your hobby, but don't know where to start?

Are you afraid to charge more for your knives?

Do you constantly get low-balled on the price of your knives?

'Bladesmith's Guide to Selling Knives' contains simple but fundamental secrets to selling your knives for profit.

Both audio and PDF versions are included.

Inside this book you will discover:

- How to **make more money** when selling knives and swords to customers (Higher prices)
- The **hidden-in-plain-sight** location that is perfect for selling knives (Gun shows)
- Your **biggest 'asset'** that you can leverage to charge higher prices for your knives, and **make an extra $50 or more** off of selling the same knife.
- 4 critical mistakes you could be making, that are **holding you back from selling your knife for what it's truly worth**
- The ideal number of knives you should bring to a knife show
- 5 online platforms where you can sell your knives

- 9 key details you need to mention when selling your knives online, that will increase the customers you get

2. Hunting Knife Template for Stock Removal

Tired of drawing plans when making a knife?

Not good at CAD or any sort of design software?

Make planning and drawing layouts a 5-second affair, by downloading this classic bowie knife design that you can print and grind on your preferred size of stock steel.

Here's what you get:

- Classic bowie knife design **you can print and paste** on stock steel and start grinding

- Remove the hassle of planning and drawing the knife layout during knife making
- Detailed plans included, <u>to ensure straight and clean grind lines</u>

3. Stock Removal Cheat Sheet

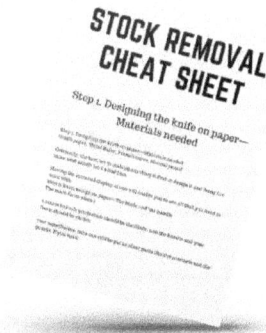

Do you need to quickly lookup the correct knife making steps, while working on a knife in your workshop?

Here's what you get:

- Make your knife through stock removal in just **14 steps**
- <u>Full stock removal process</u>, done with 1084 steel
- **Quick reference guide** you can print and place in your workshop

As mentioned above, to get access to this content, go to *https://www.elitebladesmithingmasterclass.com* and enter the e-mail where you want to receive these 3 resources.

DISCLAIMER: By signing up for the free content you are also agreeing to be added to my bladesmithing e-mail list, to which I send helpful bladesmithing tips and promotional offers.

I would suggest you download these resources before you proceed further, as they are a great supplement for this book, and have the potential to bring an improvement in your results.

CHAPTER 1: TOOLS OF THE TRADE

This book has been designed to include as little filler material as possible. This allows you to jump into a process and start working on it immediately. One way to approach this book is to read a chapter then apply it in your workshop. Try practicing and understanding the techniques in each section before you move on to the next one. This will allow you to absorb the information better and memorize one technique before you begin with the next.

But before you even begin working on metals, you do need to know all the essential tools for the process.

Tools for Your Workspace

The process of knifemaking is more than just about applying skills to create something. As you continue to make progress in knifemaking, you will be using various tools to get the results that you want. You need one set of tools to get the shape of the knife and another to make its surface smooth. The following tools will act as a starting point for making your first knife based on the processes mentioned in this book. As you get familiar with the process, I encourage you to experiment and improvise with the tools to find a technique that works best for you.

Workbench

Your workbench is going to be your main space for a lot of your work. But more than that, it will act as a space to host all the vital tools you require. One of the essential things to note is that your workbench should be elevated to a comfortable height. By doing so, you won't have to bend over and strain your back muscles while working. Additionally, it's very important that your workbench be stable so that it doesn't move around when you're working on your blade. If anchoring it to the floor or wall isn't possible, try to add weight on the bottom to create a solid base with as little movement as possible.

Angle Grinder

It is not necessary to have this tool right way. However, it definitely saves time when using it for different tasks such as grinding or cutting metal. An important point to note here is that you need to be extremely cautious when handling an angle grinder.

Drill

Since you will need a tool that can easily punch holes into the knife's tang (more on this later) so that you can attach your handle, it is important to get yourself a drill press. You can also make use of a hand drill, but a drill press allows you to make accurate holes while you balance your knife carefully.

Files

While the bulk of the work is done using power tools, it is nonetheless useful to have a file around. You can accomplish specific tasks efficiently, such as quickly removing a minor metal burr or fine-tuning the knife design. When you have a file around, you can take your time to work on your knife and get the result that you want.

Belt Grinder/Belt Sander

A valuable tool to have in your work space, and as you will notice when we start working with knives, it is an essential part of the process.

You can always use the angle grinder to perform many of the tasks of a belt grinder, but I would recommend keeping the angle grinder for cutting. The belt grinder is much safer and easier to use. Plus, you can perform a plethora of tasks on it including shaping handles, grinding bevels, adding the finishing touches to your knife, and more.

One of the cheaper options for a belt grinder is the 1 x 30-inch version. The grinder itself won't be as rugged or versatile as the other varieties that you can get in the market, but it will help you out as a beginner.

The more expensive version is the 2 x 72-inch belt grinder. With the step-up in investment, you also get a hardier and sturdier grinder. This will improve the quality of the output you are producing.

Quenchant

Quenching is a vital part of the knifemaking process. For the process, you need a container called the 'quenchant.'

During the process of quenching, there is a possibility that your blade could "flare-up," or the oil could begin flaming. For such scenarios, it is always ideal to keep the container used for quenching fireproof. Additionally, you should also make sure that your quench comes with a lid so that you can quickly cut off the supply of oxygen in case of an emergency.

Many people use an improvised container as a quench. You can do the same by using a used metal coffee container. If you are using such a container, then the idea is that you are going to work with blades that can easily fit into the container.

If heat treating 1084 steel, you can use a container of canola oil for quenching.

Other items that you can use as a quench include a 5-gallon steel bucket, a loaf pan, a heavy metal roasting pan, and even a used fire extinguisher with the top part cut off (yeah, that's a thing). But no matter what container you choose, make sure that it is fireproof and has a lid. And oh yes, do not start cutting fire extinguishers without proper equipment and knowledge. It is complex, and you might injure yourself.

Hacksaw

The hacksaw works similarly to an angle grinder, but it allows you to make fine adjustments whenever you are cutting. You can stop and adjust your position much easier with the hacksaw than with the angle grinder. It also takes longer, since you are doing the cutting by hand.

Tools for Smithing

There are tools that you will need in your workplace, and there are those tools that you need to have with you in person. They all serve an important purpose, as you will notice when we are working with knives.

Here is something to remember: the number of tools that you can find in a bladesmith's workshop and various purposes that they serve can be quite overwhelming to understand. You might visit a workshop and be amazed at some of the objects you find, wondering just what they can be used for. Since you are a beginner, you don't have to worry about too many tools. You need only to have the below for now.

Hammer

A bladesmith striking iron with a hammer is, in fact, the quintessential symbolic representation of making a powerful weapon or tool.

There is a reason for that. Your hammer is going to be of the most versatile tool that you are going to use in your knife making process. Think of the hammer as an extension of your arm, reaching out to touch the metal and work with it where your hands cannot (after all, the metals are heated to high temperatures).

Hammering is all about efficiency, so take the time to make your hammer as comfortable as possible. If you have to, you can shave the handle down to fit your hand snugly. You should be able to grasp it easily without having to use a death grip. Your body will get used to working with a particular length, and it helps you become more accurate in time.

One of the most important factors you should consider when choosing a hammer is its weight. It should be light enough that you are not going to cause muscle fatigue. The head of your hammer should weigh anywhere between 1.5 to 3 pounds.

The next thing you should focus on is the length of the handle. It should ideally be the same length as the distance from your elbow to the tips of your fingers. This allows you to work with the metal without keeping close to it.

Tongs

What the movies don't show you is that many bladesmiths use their tongs. If one hand holds the hammer, the other holds the tongs, even though tongs don't usually get the coverage that they deserve.

In a knife making process, you are usually handling metals that are heated up to 1,500°F. You can't touch metals at that temperature using your gloves. You need special equipment that can hold the metal securely and comfortably.

To put it plainly, you need tongs.

Anvil

If you are planning to buy a brand-new anvil, then you might have to shell out a little cash. But usually, you will be able to find anvils for sale or being sold second hand. You can get any type of anvil, but beware of any deep chips or indentations that will cause problems when you use it in the future.

Many bladesmiths like using anvils that are at least 100 pounds, but you can use something smaller when first starting out. The one thing I would like to point out is that the lighter the anvil, the more energy it absorbs when you are working with metal. The heavier it is, the more the metal will feel the impact. This is important because you want the metal to feel the impacts rather than the anvil.

Anvil Stand

This is not vital, but it helps you keep your anvil and the metal steady when you are working. Sometimes, you might experience situations where your anvil can slide along the floor.

The most important reason for getting a stand is to elevate the anvil. Unlike how you see them in movies, the anvil is not that tall. This means you can end up leaning or bending down to work on your metal. Even if you use a chair, you might be in an awkward position to perform your metalworking process. By using a stand, you can raise the anvil for greater comfort.

Want to know the ideal height for your anvil? Place your arms by your side and make a fist. You should place your anvil at the same level as your knuckles.

Another thing that you should focus on is the positioning of your anvil. Ideally, it should be close to your forge but not too close. You should be able to transfer the metal from the forge as quickly as possible with enough room to navigate.

Figure 2: A 55-pound anvil placed on a metal bench.

Forge

You can typically find two kinds of forges, coal-based, and propane-based. Both have their own set of advantages and limitations.

Coal-Based

These forges are quieter than their propane counterparts. You can also easily get the heat centered around a particular area. This allows them to be truly versatile.

Their drawback is that they are not ideal for beginners. They require a lot of maintenance. If you are not careful or used to them, then they can easily overheat and end up ruining your work. Also, because coal is not particularly clean to handle, you might find them getting dirty quite often.

Propane-Based

The best part about propane-based forge is that they can easily be started and require less time to get used to or work with. They are quite convenient to use and have a higher degree of portability than coal-based forges.

On the flip side, they are quite noisy and require proper ventilation. You should make sure that you are not using propane-based forge inside an enclosed space or there are risks of carbon monoxide poisoning.

The Ideal Forge

If you are getting started, then you could try using a two-brick propane forge. But as mentioned above, make sure that your workspace has proper space. If you are working inside a garage, then make sure that the garage door is wide open to carry out the exhaust and smoke from the propane forge. Additionally, if you are feeling faint or uncomfortable working with a propane forge, then make sure you stop your work, find enough ventilation, move the forge to a different location, and try again.

Starting With Steel

You are going to come across various steels to work with. There are steels such as 5160, W1, W2, O1, and more. Each type of steel that you find has its own properties.

But which one should you start with? Is there a beginner steel that you can use to practice knife making? Is there a steel that does not pose too many challenges?

Fortunately, there is.

In the world of knifemaking, 1084 is considered a beginner's steel. This steel is one of the most uncomplicated steels that you can work with at home. 1084 is part of the ten series of steels. The higher the number, the higher the percentage of carbon that they have. For example, 1045 has 0.45% carbon out of the total composition of elements in the steel. Here are the remaining steel variations of the ten series:

10 Series Steel	Percentage of Carbon
1045	0.45%
1050	0.50%
1055	0.55%
1060	0.60%
1084	0.84%
1095	0.95%

As you can see from the table above, 1084 has enough carbon to give you the sturdiness that you require for your knifemaking. At the same time, it is easier to heat treat 1084. This makes it ideal for getting used to various processes.

Additionally, it takes time to work on other forms of steel. If you start off using a more challenging type of steel and you do not like the results that come out in the end, then you are going to be that much more disappointed and completely exhausted by the entire knife making process. This is why, when you start with the 1084 steel, you won't mind making mistakes.

When you master 1084, you can feel free to move on to 1095, which has a higher carbon content and requires careful attention and skill while going through heat treatment.

CHAPTER 2: ANATOMY OF A KNIFE

Before you start working on a knife, you need to know more about its anatomy. This knowledge will help you understand just what you are working with and the parts you are going to handle.

Basic Anatomy

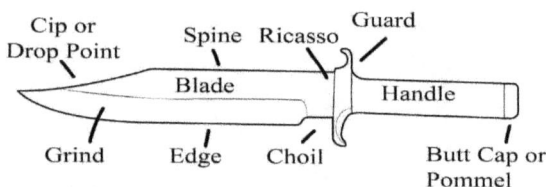

Figure 3: The anatomy of a knife

Point

The point is the tip of the knife. Or in other words, it is the business end of the knife.

Belly

The belly of the knife is the arc that is formed along the edge of the knife. It is a curved region that starts from the middle of the knife's blade and reaches the point.

Spine

This is the unsharpened top part of the knife. Essentially, it is the blunt side that allows people to use their fingers to press down on the knife.

Edge

The entire sharp part of the blade. Some knives have a single edge while others have a double edge.

Serrations

Some blades have a sawtooth-like design on the edge of the knife. These designs are often called serrations.

Blade

The blade is the sharp region of the knife that includes the edge, point, serrations, spine, and belly.

Bevel

When you look at a knife, then you will notice a slight incline that leads to the edge of the knife. This incline is called the bevel. The higher your bevels are, the more cutting power your knife has.

Tang

The back portion of the knife. Essentially, this is the 'handle' part of the knife without the actual handle attached to it.

Handle

The handle is the covering for the tang. The handle can be made out of wood or leather.

Pin

Not all knives have pins, but they can be spotted easily on the handle of the blade. They are the small spots on the blade which you add to secure the handle to the tang.

Advanced Anatomy

Ricasso

This is the thick part of the knife that lies between the blade and the handle. The ricasso is mainly used to provide extra reinforcement of the knife.

Pommel

This is the butt of the knife, and sometimes, you might find knifemakers turn it into a unique design or a cap-like formation.

Quillons

These are protrusions from the handle. You will find a pair of these, one on the pommel and one between the ricasso and the handle. They are usually created to prevent the hand from sliding up and down the handle. Can also be referred to as the guard if you see them only formed between the ricasso and the handle.

Quillons are usually designed on only one side of the knife. A guard is formed on both sides of the handle such that if you hold the knife vertically, it will look like a cross. In fact, a better way to imagine this part is to imagine a vertical knife. The horizontal section that goes across the knife is the guard.

Bolster

With some knives, you might notice a thick junction between the handle of the knife and the blade. This thick section is used to provide a smooth transition from the tang to the blade and is called bolster.

Choil

Some knives have a slight depression between the edge and the ricasso. Such a depression is known as a choil.

Knowing the Knife

When you know the different parts of the knife, then it becomes easy for you to design your knife. Do you need a pommel? Are you planning to make a double-edged knife or a single-edge? Since you are a beginner, can you make the guard easily or would you like to try making a knife without a guard first?

By recognizing different parts of the knife, you will be able to create one that fits your idea. Additionally, when you want to make changes to a specific section of a knife, then you will know the name of the part you are focusing on. This becomes important when you are trying to describe your knife to someone else.

Blade Profiles

Different knives will be better suited for various tasks, depending on their blade profile. The profile is the term used to describe the overall shape of the blade and gives the blade its look. Learning the basic blade profiles can serve as a guide while designing your blade based on the specific functions you want it to serve.

One of the most common blade profiles is the drop point, which is favored as the best style for survival and hunting blades. A drop point is characterized by the spine of the blade "dropping" down from the handle to the tip, with the tip at the center axis of the blade. The spine extends the full length of the tip, which makes it much stronger and less prone to breaking. It also makes an excellent knife for carving.

A clip point blade is another common type of blade profile and is named after the "clipped off" appearance of the tip of the blade. The tip is sharper and thinner and, therefore, more suited for stabbing and piercing. However, this thin tip also makes it much weaker than a drop point blade, and it tends to break more easily.

A tanto profile is sometimes used in a military-style design and fighting utility knives. This profile has a spine that slopes slightly down to a point that is sharp and angular. This makes for a blade with a tip that is great for piercing, stabbing, and general utility.

A spearpoint design has a point that meets in the middle of two symmetrical sides, much like the tip of a spear. A spearpoint can have either one or both edges sharpened and has a strong and sharp point. This feature makes it a great knife for piercing or stabbing. However, it's very difficult to do any fine carving or detail work with, and it isn't a practical daily-use knife.

Designing a Knife

There are a few things that you should remember when you are designing your knife.

When you are designing a knife, make sure you know what you will primarily use it for. This helps you understand how you would like to design the edge, whether you need a guard, what kind of point you should create, and other useful information. A lot of people who start making knives think that they would like to design a knife that can achieve practically every purpose. But such a knife does not exist. Designing a knife in a particular manner means that you have to make sacrifices in other areas.

To make a knife, you will require a knife template. The best way to learn how to make a template is to see it in the process. So, here are the steps to create your very own knife template. In this template, we are going to create a simple hunting knife.

On a piece of printer paper, draw two horizontal lines that are no more than 2 inches apart. Since you are getting started, use a width of 1¼ inch. Between these two lines, you have to create your knife.

The overall length of the knife should not be more than 15 inches. However, with our measurements, the knives won't even be that long.

Let's focus on the blade first. A 4-inch blade might be too long for some and anything that falls below 3¼ inches might be too short. Mainly, you should keep the blade length at 3⅞ inches or if that seems too precise, then make the blade between 3¼ and 4 inches long.

Creating the handle is tricky since different people have different hand sizes. But what you should do is form a handle that balances the blade. At this point, you should ideally be looking to create a handle that is about 4 inches long. Some knife makers can go up to 4¼ inches long for the handle, but you should remain within the 4 ¼ inch mark.

Creating a Template

Draw out your knife. This drawing will eventually be your knife, so take your time and play around with the design.

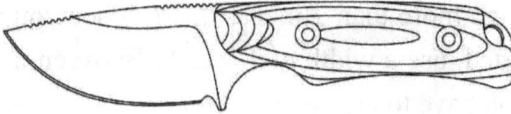

Figure 4: A simple knife template

If you want to get the printable version of this knife template for FREE, go to **https://www.elitebladesmithing-masterclass.com**, and enter your e-mail.

When you settle on a final design, make a photocopy to keep as a reference. Then, cut out the paper knife and glue it onto a piece of wood using spray adhesive.

Using a band saw, or hacksaw, carefully cut out the design onto the piece of wood. Fine-tune the template by using a wood file to get rid of the saw marks and shape your knife by taking away everything up to the edge of the drawing. The more precise you are, the more you'll get a feel for what your knife will be like.

Hold the template in your hand and get a feel for it. Does the handle need to be longer or shorter? Is the blade length what you're looking for? Are the blade length and handle length the right proportions? If you have any doubts in the design, this is the time to fix it. It takes far less time to make a new template than to try to fix design flaws in your blade as you're working.

CHAPTER 3: MAKING A KNIFE BY STOCK RE-
MOVAL

In this chapter, we are going to learn how to make a knife using a method that is ideal for beginners; stock removal.

We have already created the template for the hunting knife in the previous chapter. The topic of grinding bevels and heat treatment are explained in more detail, later in this book.

As we had mentioned before, we are going to start off by using the 1084 blade. We already know the dimensions of our knife, so it should be relatively easy to discover the dimensions of the steel as well.

For your knife, use a 1084 steel that is 9 inches long. This way, even if you decide to make the blade 4 inches and the handle 4¼ inches, you will have plenty of space to work with. The width of the metal that you choose should be no more than 2 inches.

If you find yourself in possession of steel that is longer and wider than the dimensions you have chosen, then all you have to do is use your trusty angle grinder to cut off the extra parts. To cut them off, draw the dimensions of the steel (9 inches x 2 inches) and cut along the lines. Try not to waste any of the extra metal as you could use it for making more knives. The best way to work with all the extra metal is by drawing grids of your steel dimensions. For example, if you have 1084 steel that is 10 inches long and about 6 inches wide, then technically, you can make 3 x (9 inches x 2 inches sized) blades. You will have one very narrow strip remaining in the end, which you can keep aside for future projects. You can still make knives out of the remaining metal. However, they might be narrower than the ones you are making right now.

You can even draw out the shape of your knife on the blade itself to help you understand how to get the shape that you want.

Once you have done that, follow the steps below:

1. Typically, you might find out that your steel comes with the run-of-the-mill (no pun intended) mill scale. When you use metals, they should ideally have a smoke gray color. But with a mill scale coating, you can end up having a dark gray layer. There are two ways of removing the mill scale, the mechanical method, and the chemical method.

2. In the mechanical method, you use your belt grinder to slowly chip away at the layer until you can finally see the mill scale removed. You can use a 50 or 60 grit belt for a rough grit and then move up to 100 grit belt for a final grit.

3. The chemical method is the easier method, but it takes a while to get all the coating off. First, you will need the below items:

    a. A 2-gallon bucket

    b. 2 x 5-gallon bucket

    c. Chemical resistance gloves

    d. Safety goggles

    e. White vinegar (you need enough to make sure that your steel is completely submerged into the vinegar)

4. When you are ready, take one 5-gallon bucket and then pour the white vinegar into it.

    a. Next, take the 2-gallon bucket and drill some holes at the bottom using your handheld drill.

    b. Take the piece of metal that you would like to descale and then place it inside the 5-gallon bucket that has the white vinegar in it.

    c. You now have to wait for at least 24 hours. During that time, keep turning the metal around every 4 hours so that the vinegar can reach every part of the metal.

    d. Do not dip your hands directly into the vinegar. It might not be as dangerous as other forms of acid, but it still has a burn to it. Also, make sure that you are using goggles to protect your eyes from splashes when moving around the metal.

e. As you can see, this method of descaling is rather long, but it involves less activity, and you can use it when you have a busy day ahead of you.

f. When you finally take out the steel from the bucket, you only need to hose it down with water to remove any traces of mill scale. If you find any stubborn mill scales, you can use the belt grinder, but you won't be putting a lot of effort into it. The mill scales usually come off within a minute or less of subjecting it to the grinder.

g. You can also fill the second 5-gallon bucket with water and use that to take off any traces of the white vinegar. Lift the 2-gallon bucket, wait for the white vinegar to drain, and transfer it to the 5-gallon bucket with water for a quick wash before you hose or grind of the mill scale.

h. The alternative to using vinegar is muriatic acid. However, I would not recommend using muriatic acid for beginners because the acid is highly toxic. In fact, they are so toxic that you will have to use respirators or your lungs will burn by inhaling even a small portion of the acid fumes. Plus, they are tough to handle and might need a lot of extra precautions. So in the interest of safety, let's try and avoid industrial grade acid as much as possible.

i.   If you would like to buy metal with no mill scale deposits on it, then you should ideally be looking to get cold-rolled steel. But you will be spending quite a bit to get your hands on cold-rolled steel.

5.   Once you have removed the mill scale, use a clamp to keep the steel in place. Then use your hacksaw to cut out a rough shape of the knife. You can also use an angle grinder for the job. Regardless of what tool you use, make sure that you are well protected, and you are careful when using the tools. Prudence is your best course of action, especially since you are a beginner.

6.   Your next step is to clean up the knife using your belt sander. This means that you are going to get the shape more refined and accurate. For this purpose, we are going to use a 60 grit belt, since all we want to do at this point is to refine the shape we have already created.

7.   You are now ready to add a bevel to the knife. At this point, it will be a little rough. Go ahead and mark the bevel and edge on your knife using any marker. Then you run the knife on the belt sander/grinder until you get the edge that you

want. Essentially, you need to mark the center of your knife or the point from where you edge begins. Make sure that when you are adding your bevel, you keep the part towards the center or the point of origin of the bevel thick.

8. There are 3 different types of bevels you can choose from: flat grind, convex grind, or hollow grind. The full guide to doing this is given later in the book.

9. Now that you have given the knife its shape, it is now time to drill holes in the tang. You should always do this before going into the heat treatment process. You can use any drill for creating the holes, but the drill press gives you greater control and accuracy because you can hold the knife in place while you make the holes.

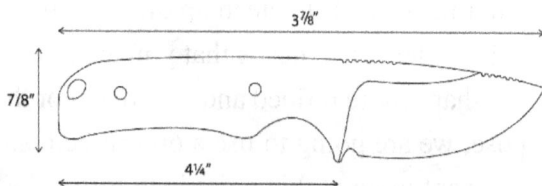

Figure 5: The knife template with measurements

10. Time to heat-treat the metal. The instructions are for 1084 steel. Place it into the forge and allow the steel to heat until it enters the yellow range (or in other words, when the metal turns bright

yellow). You are aiming to get the metal to a non-magnetic point. If you would like to check if you have successfully done this, place a magnet nearby and see if the metal attracts it. When you have reached the non-magnetic temperature, then keep the metal in the forge for about 15 minutes. Full heat treatment instructions are given in the following chapters.

11. You are now going to quench the metal. Make sure that you have canola oil that has been heated to 135°F, to use as the quenchant. At the end of those 15 minutes where you placed the metal into the forge, transfer it directly to the canola oil. As we had mentioned before, make sure you have a lid ready nearby in case you need it.

12. Once you have finished quenching the metal, take it out for the tempering process. Preheat your oven to around 400°F. Place the metal into the oven for about two hours. Take out the metal, allow it to cool. Then place it back into the oven for another two hours. You can also make use of a blowtorch if you have one. Or you can use a toaster (make sure the entire blade fits into the toaster).

13. After completing the tempering process, it is time to remove the scales from heat heat treatment and finalize the bevels. Head back to your grinder and then add the finishing touches. Grind down the

knife to the thickness that you want. You will notice that at this point, your knife looks more and
more like the knife that you had in mind.

14. Your knife is now ready for the gluing process.
To glue the handles on the knife, use epoxy to get
the job done. One of the popular epoxies that you
can find in the market is T-88, but you are wel-
come to use any brand that you are comfortable
using.

15. Shape your handle using the belt sander.

16. Polish your blade using sandpaper.

17. Finish it off by sharpening it.

A Couple of Tips to Remember

- You can also drill holes into the knife design that
you made on the steel bar before using a hacksaw
to cut it. This is because it is much easier to clamp
down a rectangular piece of steel, as compared to
a knife blank.

- When you are working on the belt sander, make
sure that your tool rest is as close to the belt as
possible. If it is not, then there is a possibility that
the belt with catch the knife, wrestle it in between, and harm your fingers in the process.

CHAPTER 4: FORGING A KNIFE
(FULL TANG KNIFE)

Before we go into the actual process of forging, it is better to understand a bit about the process.

After all, you have your workspace all set up, your design laid out, and a piece of steel ready to go. At this point, you should make sure all your tools are where you can have easy access to them, place your blade in the forge, and get it cranking.

To properly work your blade, you'll need to take it out of the forge when it's at an appropriate temperature. While it's possible to buy a thermometer for a propane forge, most people gauge the right temperature and the degree of workability of the metal by its color.

Here is a table to help you understand the color and the temperature of the metal when it attains that color. Use this as a reference whenever you work with your metal.

Fahrenheit	The Color of the Steel	Process
2,000°	Bright Yellow	Forging
1,900°	Dark Yellow	Forging
1,800°	Orange Yellow	Forging

1,700°	Orange	Forging
1,600°	Orange Red	Forging
1,500°	Bright Red	Forging
1,400°	Red	Forging
1,300°	Medium Red	-
1,200°	Dull Red	-
1,100°	Slight Red	-
1,000°	Mostly Grey	-
800°	Dark Grey	Tempering
575°	Blue	Tempering
540°	Dark Purple	Tempering
520°	Purple	Tempering
480°	Brown	Tempering
445°	Light Straw	Tempering

Remember that when you are using the table, you don't have to get the exact shade of the color mentioned above. Typically, if you get an orange-yellow or orange shade, then you have reached the orange temperature range. Which is why you might always hear bladesmiths mention that they heat the metal to a particular range.

We are going to use the same knife design that we created in Chapter 2. If you haven't designed the knife already, go ahead and create the design now.

Once again, we are going to use a steel that is 9 inches long and 2 inches wide. This time, however, we are going to use 1095 high carbon steel.

We are going to begin by de-scaling the steel. The method to de-scale is taught in Chapter 3. After you have completed de-scaling the steel, here are the steps that you need to take to forge it.

Here are a couple of things to remember:

- Steel needs to undergo changes during heating to become malleable (or flexible in layman terms, but we do not use the term flexible here as it might indicate the metal can be stretched). If you try to hit a piece of steel that's too cold, you'll only be working the outer layers of the metal, and you'll get a unique effect that looks like mushroom shapes taking place on the metal.

- It's also possible to get too high of a temperature, which is most likely when using a coal forge. As a beginner, you should start with the propane forge but if you are already using the coal forge, then make sure that you are careful and not overheating the metal. Here is a tip for you to follow when you are using a coal-based forge: take care to arrange your coals so that you can lay your steel across the fire evenly. If you concentrate the heat towards the knife to a particular area, then it is possible to melt the tip of your knife entirely off. Or an easier method is to, of course, use a propane forge!

A Game Plan

Before you start hammering, you need to come up with a game plan. Your steel will begin to cool down immediately when you take it out of the forge, and every time you place it on the anvil, it rapidly draws heat away through conduction. This means you only have about six to eight good hammer strikes before you need to put it back in again to heat up. So if you are in the process of shaping the metal, then you have to be precise with your strikes. Use your time wisely! Take a good look at your steel and plan exactly where you'll be hitting it before you put it back in the forge. Make a mental plan of the steps you'll be taking so that you don't have to waste any time trying to figure that out when your blade is heated and ready to work. If you do have to stop hammering for a moment, hold the steel up instead of letting it rest on the anvil to prevent unnecessary cooling.

As a beginner, your ability to make accurate strikes on the metal is unlikely. You might not be able to get the shape that you want quickly. But that is okay. Take as much time as you need and place the metal into the forge through multiple cycles to form the shape you had in your mind.

There are no mistakes here, merely lessons to be learned while you work the forge.

Working With Steel

It's useful to think of the hot steel as clay. Imagine hitting clay with your hammer and what would happen to it as you delivered blows. While steel will be a lot harder to move than clay, the basic principles are the same. As you apply force, the malleable steel will move away from that force in the direction of least resistance. As you work, you will be adjusting where you are striking with your hammer as well as the direction of your strike. In this manner, you can control not only the force but where the steel will have a tendency to move by manipulating the path of least resistance.

As you work, the steel will eventually bend up at the ends, losing its flatness. It seems like common sense to want to hammer on the upturned ends to flatten the steel again, but this doesn't allow you to apply the right kind of force. By flipping the steel over and striking it in the middle, both ends will be pushed flat against the anvil as the middle moves away from the force of your hammer.

Keep an eye on the steel as you work. Use the last strike or two to ensure your steel stays as flat as possible.

Now we are going to see how you can use the forge to give shape to the knife.

1. The first thing you are going to do is figure out where the knife's tip is going to be. Since we are not going to be using the hacksaw and grinding the knife, we are going to be using the forge to shape out the knife.

2. Heat the steel until it reaches the yellow temperature range.

3. Once it reaches the temperature, use your tongs to take out the steel. Now you are going to hold the steel on the anvil. Hold it in a way that the future cutting edge of the knife is facing down.

4. Hit the top corner at a 45° angle. This is because we are going to start with the knife's drop point, which is an easy way to create the knife point. As you strike, you'll notice the steel mushrooming out. When you see this happen, then place the steel on its sides and hammer it until the mushroom shape disappears.

Figure 6: Forging the right knife is all about patience and careful understanding of the technique.

1. Go back to the previous position and continue hammering the steel at a 45° angle.

2. Keep repeating this process until you have what looks like the profile of the knife tip pounded in.

3. Currently, the point or tip of the knife will be positioned more towards where the edge of the knife should be. This tip will change position as you continue to move the metal, moving from the bottom of the blade to the top. The natural tendency of the metal is to push things away as it gets thinner, which is why, as you begin working on the edge, this tip or point will end up rising to the top of the blade, closer to the blade's spine level.

So if you start getting worried about why your point is not in the correct position, don't! We still have a lot of steps to go.

4. So back to the knife. We are now going to work on the knife's edge.

5. Look at your steel again and decide how long you want the cutting edge of your knife to be. This is based on the total length of the knife's blade. So you should have placed it anywhere between 3¼ inches and 4 inches long.

6. Once you have decided the length of the knife, heat it again until it reaches the yellow temperature range.

7. Here is a trick that you can use to make a mark on the blade so that you will know where the blade of the knife begins. Hold the blade over the edge of the anvil and hit one side of it (the spindle side of the knife) of it with the hammer to make a small indented mark on the cutting edge side of the blade. This mark will let you know where the handle of the blade starts from (make sure that you are using your measurements to decide where this mark should be). You can choose to remove this mark later during the forging process (it happens as you work on the metal) or alternatively, many bladesmiths make the indentation deeper and make a choil out of it. For now, we

are going to focus on how to make a basic knife without the choil.

8. Now that you know the edge. It is time to hammer it out and make it look sharp!

9. **Tip:** Use a little bit of water on the anvil. This will help blow out the knife scale, with each hammer blow.

10. Holding your steel flat on the anvil, pound along the edge. Flip the metal and hammer on the same part from the other side. This will cause the metal here to thin out and will begin to create the bevels of your knife. Notice how, as you thin the edge, the tip will slowly start moving up towards the spine side of the knife. This means that you are going to get the point to the proper place.

11. As you make progress with your blade, keep an eye on its overall shape. In addition to maintaining overall flatness, you'll have to work to keep the spine of your blade straight as well. As you notice it losing its straightness, flip the blade onto its edge so that the spine is on the anvil. Now hammer the spine until you see it becoming straight. But won't this affect the edge as well? Of course, it will! In which case, flip the knife onto its side and hammer the edge back to its correct shape. You need to constantly correct the shapes of either parts of the blade.

12. As you keep working on the metal, you will notice that it will get a rough shape. You don't have to work until the edge is truly sharp. All you are doing is getting a rough outline of the final knife that you are going to work on.

13. Once you get a rough shape, shift your attention to the handle area, or tang. Hold the blade on its edge, with the spine side on the anvil.

14. Start hammering the knife from the part where your indentation starts (the one you used to mark the start of the blade). Once again, hammer the side of the blade to remove the mushrooming effect that the metal gains during the process. Before working on the tang, the edge and the tang will look like they are connected. Once you start hammering the knife, the tang will form a handle shape, the metal getting narrower.

15. By now, you should be able to see the full shape of your knife starting to form. Use the techniques you've learned to try to focus on areas that show mistakes and further refine the profile of your knife. There's no substitution for time and learning from your mistakes, so don't be afraid if you make them. Every mistake is an opportunity to learn.

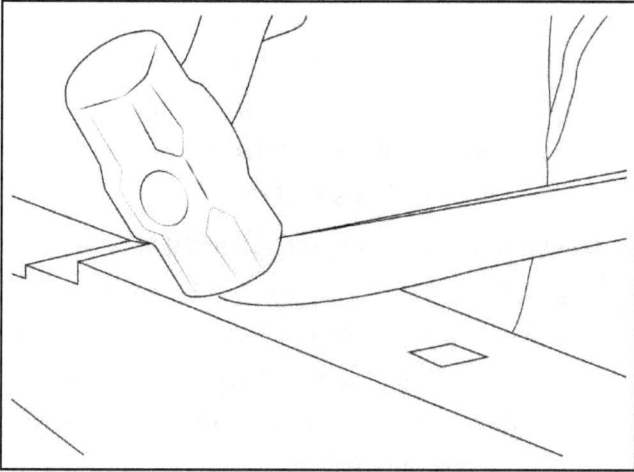

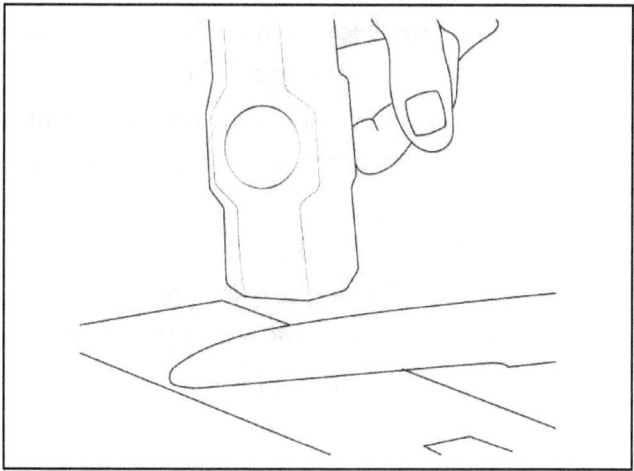

Figure 7 & 8: Hit the steel at the corners
to develop the point

16. Once you have completed the profile of your
knife and you are satisfied with the results, let's
move on to making holes in the tang.

17. You have already seen how you can make holes using the drill. But since your knife is hot, how can you create these holes right now? What you do is use a technique called the hot punch. How does it work? Let's find out.

18. The first thing that you are going to want to do is get yourself a hot punch, which is like a pointed and elongated piece of steel. Look at the pointed end of the hot punch and check out how narrow or broad it is. This will let you know how big the hole on the tang will be.

19. When you are ready, heat the knife's tang until it reaches the yellow range in temperature. Once the temperature is right, place the knife on the anvil. Focus on where you would like to create a hole in the tang. Place the hot punch on top of that position and strike it a few times. Flip the knife over with the tongs, and you will be able to notice a small dark spot where the punch struck the knife on the other side. Place the punch over this dark spot and hit it with the hammer. Continue this process until you literally 'punch' out the piece of metal from the tang and create a hole. Make sure to take the hot punch out quickly after hammering, so that it does not get stuck in the hole and join with the knife.

20. Repeat this step with the rest of the holes that you would like to create on the knife.

21. At this point, you might notice that your knife might have some scaling on it. Head over to the grinder and remove the scales (you can also use the white vinegar method, but at this point, the scales won't be too challenging to remove).

22. Once we have accomplished all of that, it is time to move on to the next step: grinding bevels!

CHAPTER 5: GRINDING GOOD
BEVEL LINES

Now that making the knife blank is done, it's time to start grinding. After the profile of your knife is refined, you'll be removing layers of knife material to create bevels. These will form an angle that makes the cutting edge of your blade.

You could say that this is the defining moment in which you turn your piece of steel into a knife. Does that sound exciting? Well, let's get started. But before that, we should understand a little more about grinding.

There's a variety of tools and techniques you can use to create these grinds, and your choice will depend on your own experience and preference. Professional knifemakers make this step look easy, but it takes a lot of practice to develop their level of comfort and skill. The trick is to go slow, be patient with yourself, and put in plenty of time behind the grinder.

Before you start making the angles, make sure your blade profile is all set. You can remove any large pieces of steel outside your design with a hacksaw, band saw, or the cutting wheel of an angle grinder. If you worked your blade at the forge, then revisit the template and redraw your bevel line. Use a belt grinder, files, or angle grinder to remove all material that won't be part of your knife's final shape.

A quality grind on a knife is produced not only with great technique but with a basic understanding of blade geometry. A good blade has an appropriate balance between overall strength, sharpness, and edge retention suited for its intended use. Unfortunately, there's no one-size-fits-all knife grind. Understanding what factors affect the performance of your blade will help you to choose the best grind for your blade and allow you to have a specific goal in mind.

The first thing that you will notice is that there are different types of grinds that you can use for your blade. Here are the most popular ones:

Full Flat Grind

This grind is done in a V-shaped and works consistently from the spine to the edge. It creates a good balance of cutting ability and strength. While it is a very sharp grind, it can dull quickly but is easy to sharpen. This design is standard in kitchen knives.

Scandinavian Grind

Often used when making bushcraft knives, the Scandi-navian grind - or Scandi grind for short - is a flat grind that starts below the halfway point of the blade. By leaving a lot of material in the spine of the blade, this grind can maximize the durability of your knife. The lack of a secondary bevel means that the low angle will create a sharp edge. While the edge is not as tough as other grinds that offer a secondary bevel, it does make it very easy to sharpen in the field, even for a beginner. It is an excellent grind for carving.

The location of the bevel makes it easy to see what you're doing, and a blade with a Scandi grind will be able to cut through most pieces of wood with relative ease.

Scandi

Sabre Grind

The sabre grind is a flat grind that starts halfway up the blade. While it isn't quite as good at carving, it tends to slice slightly better. Unlike the Scandi, the sabre grind typically has a secondary bevel.

Sabre
Grind

Hollow Grind

In this grind, the bevels curve in to form a thin, very sharp edge. This edge tends not to be as durable as some other grinds, and it tends to need a lot of retouching to stay sharp. This edge can be slightly more challenging to grind as a knifemaker, but the edge created isn't difficult to re-sharpen.

The incredible sharp edge of a hollow grind makes it the grind of choice for straight razors and hunting knives. It tends to bind up at the top of the hollow when slicing through materials such as cardboard and isn't as well suited to being a utility knife as some other grind styles.

**Full
Hollow
Grind**

Convex Grind

A convex grind is a rounded grind that focuses on the edge. The mass behind the edge increases the durability of the edge, and it can be quite sharp. This grind is often used in axes, machetes, or choppers.

Convex
Grind

Chisel Grind

In a chisel grind, one side of the cutting edge has a flat grind, while the other has no bevel ground in. Because of the shallow blade angle, a chisel grind makes an incredibly sharp edge. This sharp angle also means the edge doesn't have the best durability and needs to be continuously maintained. Chisel grinds are commonly used in food preparation as well as woodworking, as the bevel makes it easy to follow the wood grain. It can be slightly inaccurate when slicing, due to the edge being off-center. Knives made with this grind are often either right-handed blades or left-handed blades, depending on which side the bevel is on.

Chisel Grind

Figures 9 to 13: Different grinds and their profiles

A Note Before Grinding

When you spend more time grinding, you develop an instinctual understanding of how your body needs to move to get the results you want. There is definitely a learning curve, so be patient with yourself. With a little bit more time, you won't need to think about your movements as much as when you're first starting.

Developing your grinding style is about consistency, so eliminate discomforts. Stand with a slightly wide-set but comfortable stance to give yourself a stable base. Keep your elbows tucked against your sides and lock them into your hips. Instead of using your arms to move the blade, move from your core. Shift your weight steadily in your hips and think about using controlled and calculated movements. By working to create a pattern in your movement, you will find a comfortable rhythm that will make grinding much more predictable.

Finally, make sure that you do not take any substantial risks. Be patient with your work and make sure that you are comfortable understanding the basics of grinding.

Creating a Grind

Let us work with the Scandi grind technique and then you can use the same ideas for the other grinds:

- The first thing that you need to do is create the bevel outline in the metal. To do this, take out a permanent marker and then find the center line of your cutting edge.
- Mark that outline. If you like, you can even color the entire edge of the blade using the marker.

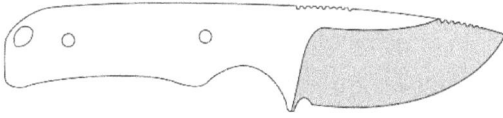

Figure 14: Mark the entire area you want to grind

- The next part is a little tricky, so make sure you read this instruction carefully before putting it to practice. Take a hand drill (or scribe) close to the outline of your blade (near the center) and run it along the flat surface, dragging the tip along the blade edge.

  o Let's imagine that you are creating a knife that is 2 inches wide. You have decided to draw the outline at the 1-inch mark.

  o You are going to use the drill to create a line just below the outline (this process is called scribing), allowing you to see where you would like the bevels to meet. Flip the blade and do the same thing again.

  o Now you can mark the bevel outline on the blade.

  o Alternatively, you can use only the outline for the purpose, but by creating a mark, you have a better idea of where you should start the bevel.

- o This step makes sure that the grinds will be symmetrical, and more resistant to warping after heat treatment.

- We are now going to take the knife to the grinder. If your grinder does not have a fresh belt, then it might be time to get one. This is because old belts heat up faster. However, you can still work using an old belt as well as it will teach you the effects of overheating, and when you should take the blade away from the belt. As for the belt grit, you can use a 50 grit belt for this purpose.

- Bring the steel toward the belt. Gently let the steel find the spot you've created for the edge and start moving the steel gently sideways. You don't need to place a ton of pressure on the blade; move it across and let the belt do its job.

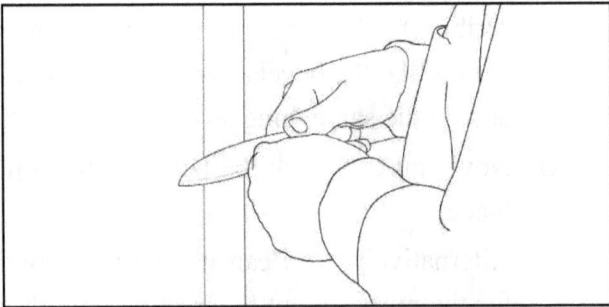

Figure 15: Sideways grinding

- Every time you take your steel off the grinder, assess how much material you need to take off and repeat the process. You should always have a clear idea of this before you grind again. Many knifemakers make the mistake of jumping back to grinding without a second thought, and end up either taking out too much material, or messing up the grind lines.

- Do not focus on one side only. Keep flipping the knife over to keep the grind lines even.

- Allow the grind to pass across the full length of the cutting edge, from the tip to just before the where the outline ends and the tang begins.

- Switch sides every few passes to keep the grind lines even.

  o Some knifemakers start at the end of the outline and work their way toward the tip of the knife. Other makers do the complete opposite; starting from the tip and moving along the knife towards the tang. There is no right or wrong way here.

  o Try working the blade both ways and see which you prefer. Remember, you are allowed to make as many mistakes as possible. Once you find the method that is comfortable for you, it will become easier for you in future grinds. But the only way you are going to find out the comfort point is by actually experimenting with different grinds!

- Another important tip to remember at this point: keep even pressure on the blade, and keep the steel moving. Check your progress every few passes. But do not be tempted to stop and check every few seconds, as this might cause the grinding process to give you a choppy line.

- Continue working your grind higher and higher up toward the spine. Every pass should be slightly higher than the last. Your grind lines should be as straight as possible.

- If you notice your line becoming wavy, try and check the amount of pressure you're putting on the blade and try to keep it consistent.

- If you find a specific area has less material taken off, try slowing down on those high spots and putting more pressure on the other side of the blade.

- Work on the blade slowly so that you can keep things even on both sides of the knife.

- Here is a tip you can use if you don't mind spending a little for different grits for your grinder.
    - Start with a 50 grit belt and complete the grinding process until the previous step.
    - Once done, switch out your 50 grit belt for a 120 grit and color in the ground surface with

a permanent marker again. Take your blade back to the grinder and start to clean up your grinds with the finer grit.

o   Keep working until the marker is completely removed and then repeat the process with a 220 grit.

o   This process of cleaning up the grind isn't only about the aesthetic of the blade but is a precautionary measure taken to prevent the blade from cracking or warping in the heat treat.

o   Any deep grooves, scratches, or sharp edges will be susceptible to cracking in the quench due to the stress created by this process. As a bonus, this will likely make the blade easier to clean up after the heat treatment process.

The Hollow Ground Edge

The hollow ground edge has a concave edge. This form of an edge is well suited to blades that will be mainly used for slicing. Examples of such blades include skinners, hunters, filet knives, etc.

The main reason why the hollow ground edge is suitable for such knives is the fact that it produces a very thin edge that can be sharpened quite easily. However, because of this thin edge, the blade can be somewhat fragile compared to other forms of grinds. This is why it is not prudent to make a hollow ground edge if you are going to be using your blade against heavier substances such as bone, wood, or materials with similar thickness. An important fact to know here is most of the blades produced around the world today are hollow ground. It could be because not many people are looking to cut bone or thicker materials!

Here is how you can achieve a hollow grind. Follow the steps in the process for the Scandi edge until you come to the part where you begin grinding. The process will be a little different for the Scandi grind.

- Take the blade and slowly bring the edge to the surface of the wheel of the belt.
- Now start the wheel of the belt and allow for the grind to form. Flip the knife over and then work on the other side. That is essentially the basics of the hollow grind edge. For beginners, getting the perfect grind might not be easy. However, with practice, you should be able to get the edge that you require.

The Flat Grind

The flat grind creates a nice balance between the hollow grind and the convex grind. One of the main advantages it provides is that, since it draws from both the hollow grind and the convex grind, it has an excellent edge that can bear the brunt of heavy chopping. Additionally, even after multiple chopping sessions, it can still retain its sharpness.

- The flat grind is similar to the hollow grind but is simpler to perform. This is because you are not focusing on the edge alone but on the entire blade.

- Your technique involves a process similar to the hollow grind. Bring the edge close to the grinding wheel or belt.

- When the edge is sharp, continue working on the blade towards the spine.

- Once you are done, you should notice a linear slope that starts from the edge and goes all the way to the spine.

The Convex Grind

A convex edge is where the blade has a bevel on each side that is slightly curved. Convex edges are said to be rather difficult to accomplish by hand. The ideal way to work on these grinds is by using a belt grinder. Here is how you can create this grind on your knife.

- For convex edges, we are first going to make sure that we leave off a little bit of thickness on the edge because we do not want it to get too thin.

- You want to start off by first flat-grinding the edge.

- Once you have completed the flat grind, you should then use a 60 grit belt on the grinder. You need to hold the knife at a slight angle, but not too much because you need the sides of the blade to touch the belt.

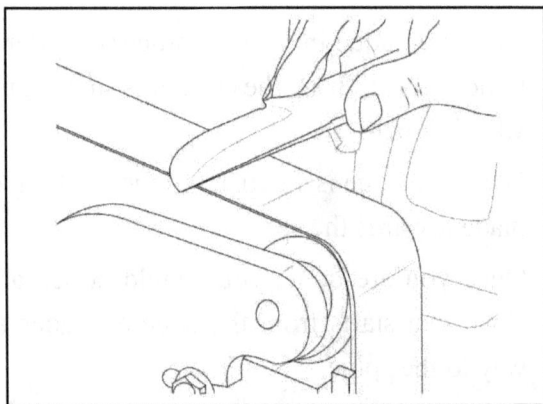

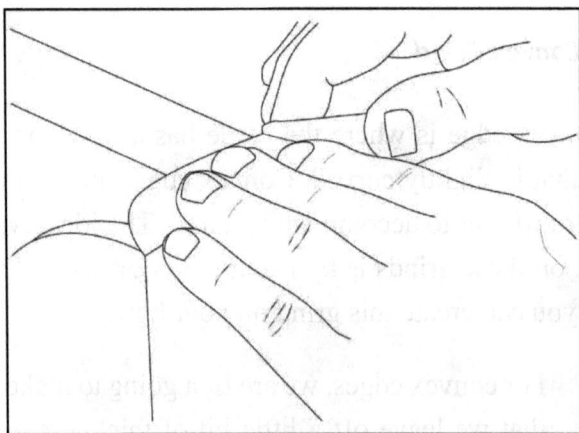

Figure 16 & 17: Grinding on top of the
belt sander helps

- With that angle, bring the side of the blade to the belt and then start grinding.

- While grinding, you need to move the knife back and forth a little bit along the edge. Start doing this on one side and then flip the knife over. Continue grinding the knife on the other side at an angle.

- Flip the knife over every two rotations of the grind.

- Once you have completed the grind, you now have a convex shape along the side of the knife.

- Now we focus on the edge. Touch the edge lightly to the belt. Don't press down too much on it.

- Move the knife and allow the belt to form a nice convex edge.

- Flip the knife over and work on the edge from the other side.

- Eventually, you are going to achieve a nice convex edge.

Tips for Grinding

- Always remember to use your hips instead of your wrist while shifting grind lines. Aim to keep your elbows close to your sides, your shoulders back and your stomach tight. All of these simple adjustments will allow you to have better control over your grinding process.

- Be confident near the grinder. Keep your movements controlled and stable. As we had already seen, do not try to press against the grinder too hard.

How Thin Should The Edge Be?

Here is a common question that many knifemakers come across: how thin or thick should the knife-edge be? What you should look for is the purpose of the knife.

The general rule of thumb is that thicker blades are used to cut harder materials. You might use them for cutting wood or skinning game. On the other hand, a thinner knife is used for slicing, like the kitchen knives.

So how thick or thin should your knife edge be? Well, if you are using it for hunting purposes, then you should make it about 1.5mm thick. If you intend to slice through things and need the right sharpness for it, then your knife edge should be about 0.3mm.

CHAPTER 6: HEAT TREATING

Essentially, no material or finished product can be manufactured without sending it through the process of heat treating. In this process, a particular metal is heated to a high temperature and then cooled under specific conditions to improve its characteristics, stability, and performance.

Through heat treatment, you can soften a metal, which allows the metal to become more flexible. You can also use heat treatment to harden metals, ensuring that their strength is improved.

Heat treatment is essential if you are in the business of manufacturing parts for automobiles, aircrafts, computers, heavy machinery, and tools. In other words, if you want something important built, then you need to subject the material to heat treatment.

Iron, and more specifically, steel, are the most common materials that go through heat treatment. However, that does not mean that other materials cannot be treated with heat. By other materials, we mean your knife.

In short, this process is quite important, and you are going to learn to use it. But more importantly, let us look at each process and try to understand what it means.

Tempering

One of the processes of heat treatment is called tempering. In this process, you are basically altering the mechanical attributes (usually the flexibility and strength) of steel or products and items made from steel. Tempering releases the carbon molecules confined in the steel to diffuse from martensite. Martensite is a form of a crystalline structure consisting of brittle carbon that exists in hardened steel. Because of martensite's features, the steel may be hard, but it also becomes brittle, rendering it useless in most applications.

Tempering allows the internal stresses that may have been formed due to past uses to be discharged from the steel. This results in the alloy becoming more durable.

So how does one temper their steel? Firstly, the steel is heated to a high temperature, but it is not allowed to heat up beyond its melting point. Once that is done, it is then cooled in air. There is no fixed temperature for all forms of steel. They each have their own temperature range that must be reached first.

When you temper the steel, it is important to heat it gradually until it reaches the temperature you would like to work with. This prevents the metal from cracking.

Annealing

Annealing is another process of heat treatment, focused on softening the steel or reducing the hardness of the material. This is done so that it is easier to machine the steel.

In this process, the metal is heated to a temperate where it is possible to attain re-crystallization. This means that new non-deformed grains take over the positions of the deformed grains. And what exactly are grains? In metallurgy, each grain is a single crystal that consists of a specific arrangement of atoms. When you have deformed grains, then you cannot work on the metal without causing more deformity. In this case, the deformity appears in the form of cracks. When you perform the annealing process, you are forming new grains, which means you are allowing yourself to work on the metal again.

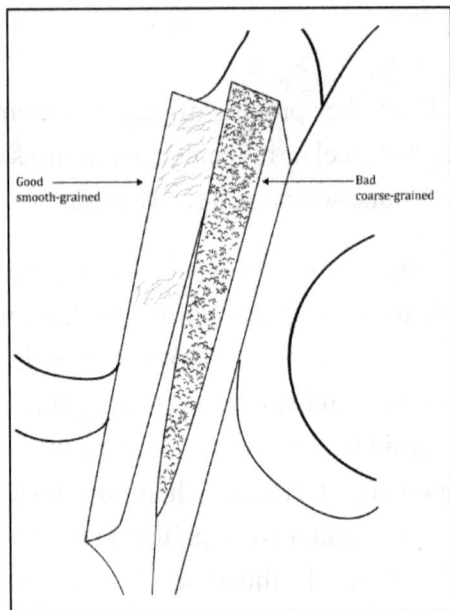

Figure 18: Good Grain vs Bad Grain

Normalizing

During normalizing, you are refining the size of the grain in the metal. After normalizing, the mechanical properties of the metal are improved.

Normalizing sets a uniformity to the structure of grains in the metal. After you have achieved uniformity, you have reduced the degree of deformity of the metal. This allows you to get a smooth finish and a wonderful product in the end.

Normalizing is usually used to remove the stresses built up inside a piece of steel and bring it back to its initial state.

In the process of normalizing, steel is heated to a high temperature and then cooled by leaving the metal at room temperature. This process of rapidly heating the metal and then slowly cooling it down makes changes in the microstructure of the alloy, making it elastic and durable. Normalizing is almost like a process of correction. This is because it is typically used when some other process unintentionally increases the hardness but decreases the malleability of the metal. What makes normalizing different from other methods such as annealing is that it uses room temperature to cool down the metal, rather than any medium or special technique.

Heat Treatment for 1084 steel

1084 has a somewhat higher manganese composition than other carbon steels in the 10XX category. Because it is a relatively easy steel to work with, it makes 1084 an ideal steel for beginners who want to start their bladesmithing adventure. It gives you enough room to make errors when it comes to heat treatment. It is known to form an almost complete 'pearlite' structure when you subject it to annealing and normalizing processes. Pearlite is a structure that features alternating layers.

Additionally, 1084 contains nearly 0.84% carbon (which is represented by the 84 in 1084) and is known to produce a good quality knife with a nice edge. Below is the full working sequence for 1084 steel.

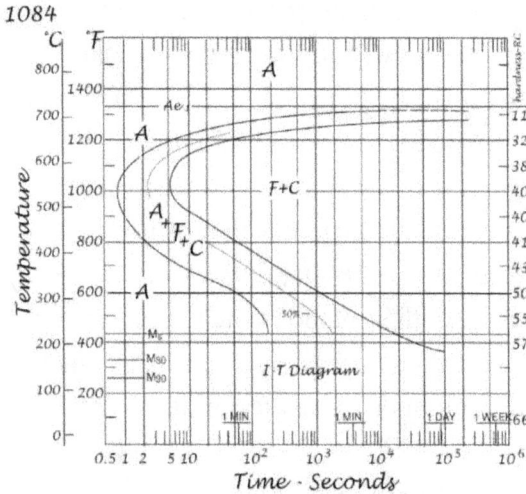

Figure 19: TTT graph of 1084 steel

Forging

You start by forging using the steps mentioned in Chapter 4. Once you have forged the knife, you can then move on to the heat treatment, beginning with the annealing process.

Normalizing

For the normalizing process, you heat the metal to 1600°F in a forge. Do not attempt to work on the metal below 1500°F. Once the temperature has been reached, soak the metal at the same temperature for about four minutes.

After four minutes, allow the metal to cool in still air. When you normalize the steel, you are resetting the crystalline structure. Through this reset, you are distributing the carbides in such a manner that they become uniform.

When you are working with steel, having an uneven structure affects its quality. Which is why, if you do not reset the structure, the carbides tend to group together tightly. Due to this, the steel will not have the sharp uniform edge that it could have had.

Annealing

In the annealing process, you start by heating the metal to 1500°F. Then you have to cool the metal, but you should avoid cooling it too quickly. You have to ensure that the metal cools at a rate of 50°F per hour or lower. I would not recommend going below 45°F for this purpose.

Pro Tip: In many cases, knife makers use an overnight cooling strategy. For this, you heat the metal to the required temperature of 1500°F at the end of the day. Ensure that the last heat of the day is slowly disappearing when you remove the metal from the forge. Once that is done, you then cool the metal in the forge overnight. This becomes handy when you have to perform other work, or you might be engaged in the evening.

At this point, you can perform your machining or grinding process, should you wish to.

Hardening

For this, you heat the steel to 1500°F. Or you can aim to push it past its non-magnetic limit. In this case, that limit is around 1425°F.

When you are working in the forge, you have to heat the metal until the metal does not attract a magnet to itself. When you have reached such a state, you heat it to a slightly higher temperature. This is just to make sure that you have truly pushed the steel into the non-magnetic area.

If you overheat the steel by keeping it at temperatures of 1550°F or beyond and you quench the metal, the metal could form grains.

To understand why grains cause harm, it is important to first understand more about grains themselves.

We all know chemistry 101; all metals are made up of atoms. Why is this important? Well, when you take a metal, then they are made up of tiny crystals of different orientations, based on the metal that you are using. These crystals are what you call grains. When you examine a single grain, then you will notice that the atoms are arranged in a particular orientation. This particular orientation can be found in every single grain of that metal.

Initially, grains do not cause any problems. However, grains tend to increase and when they do, they begin to affect the toughness of the blade. Bigger grains promote a brittle foundation, creating a knife that you might not be too proud of.

Therefore, the best way to complete this process is by heating it to its non-metallic temperature. Then keep it in the forge at that temperature for about a minute. Then remove the steel and quench it. Certain areas of the steel might only require about 1 or 2 seconds of cooling. However, that does not mean that you have to take the steel out of the forge and quickly dip in it liquid. Do not do that! Trust me that is a safety hazard. Think of it this way.

You take the metal out. You are in such a hurry to beat the 2-second mark that you knock off the oil to the ground. The metal drops, and there is a pretty big flare.

That flare catches nearby furniture or object that is flammable. Well, you know the rest.

Do not be in a hurry. The steel will hold on to some of the heat and survive for a few seconds when introduced to the air. Take it carefully and place it into the liquid for quenching. Be ready to face a small flare-up along with a high level of smoke.

Quenching

1084 doesn't need a fast quenching oil. You can use canola oil for quenching 1084. Preheat the canola oil to about 135°F. Once done, quench the metal for about 10-15 seconds.

Tempering

If you have been following the instructions, then your steel should be around 65RC. At this level, it is fairly fragile, so do not drop it. It might shatter upon hitting the ground.

Tempering Temperature		Rockwell Hardness
ºC	ºF	HRC
149	300	65
177	350	63-64
204	400	60-61
232	450	57-58
260	500	55-56
288	550	53-54
316	600	52-53
343	650	50

Rockwell Hardness Scale for 1084 steel

We need to bring the hardness of the steel down to about 59 HRC. Bring the steel to room temperature and begin tempering it once it reaches that temperature. Heat the steel to a little bit above 400°F. Temper twice. Each tempering process should be done for two hours. Allow the steel to return to room temperature between the two processes. Ideally, your method should follow this sequence: temper for two hours, then return to room temperature, and then back to tempering.

Heat Treatment for 1095 steel

Working with 1095 steel is pretty simple. It is a steel with a high carbon content, and you can use it to forge shapes easily. It does have lower traces of manganese than other steel that are part of the 10XX series (such as the 1080 steel.) However, the comparatively higher rate of carbon means that it provides more carbide that can be used for providing resistance to abrasions. However, this also means that because of the extra carbon, you might have to put in more care during the heat treatment.

If you are going to heat treat 1095 steel, I would suggest that you have a temperature controlled forge.

But let's go through all the steps in the process so that you can understand what is happening. Given below is the total working sequence for 1095 steel.

1095

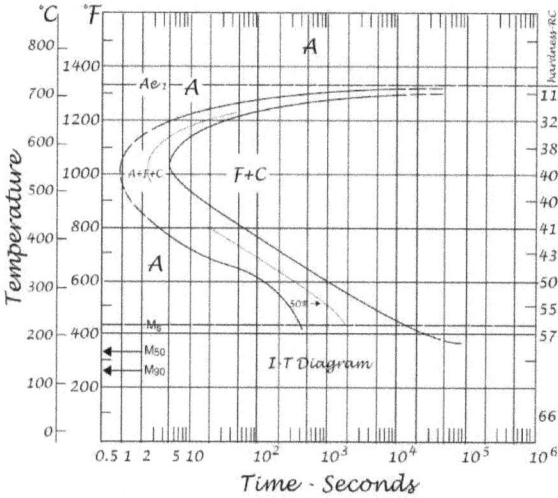

Figure 20: TTT graph of 1095 steel

Forging

We first start with the forging process. Take 1095 through the forging process that was mentioned earlier.

Normalizing

To normalize the steel, you have to bring the temperature of the metal to 1575°F. Let the metal sit inside the forge at that precise temperature, for 5 minutes. Once the 5 minutes are over, allow the metal to cool in the air till it reaches room temperature.

Another way to normalize is a little bit tricky. Get to 1575°F and choke your forge so it spits flames out of the opening and turn down the gas. It should maintain the same color inside as it did when you reached 1575°F.

Annealing

For the annealing process, you start by heating the metal to 1475°F. Cooling should be at a rate no faster than 50°F per hour.

The easiest way to cool the steel is by placing the blade inside a container of insulating, fireproof material. Wood ash is easily obtainable and a great insulator. Another option frequently used by knifemakers is vermiculite. Vermiculite is a mineral often used in gardening and can be found at any store that sells gardening supplies.

You could also go with the suggestion to cool it overnight. You have to keep the metal inside the forge to ensure the cooling is complete.

At this point, you can perform your machining or grinding process.

Grinding and Machining the Steel

You can use any of the grinding techniques that were mentioned in the previous chapter at this point.

Hardening

Heat to 1475°F which is the non-magnetic level of the knife. You can also heat just past that temperature, but ideally, stick to not going beyond that temperature. As explained before, this non-magnetic range of the temperature means that a magnet doesn't stick to the metal. Do not overheat the steel beyond 1550°F range. After heating move on to the quenching process.

Quenching

1095 steel requires a fast quenching oil. For this reason, the safest option that you can pick for 1095 steel is a special quenching oil. One of the more common oils in the market that I recommend is the Parks 50 quench oil. Parks 50 is a fast quench and is almost as fast as water. For this reason, make sure you don't grind your knife too thin before heat treatment. You start off by first preheating the oil to 70-120°F. Put the blade into the oil for about 7-9 seconds, until you notice that the hissing and bubbling subsides. Once done, take out the knife from the oil. You can also make use of quenching oil manufactured by Maxim Oil.

Finally, you can use canola oil, but that's only if you can't procure Parks 50. I only recommend using it on thin 1095 stock, up to about 1/8 inch. You could do it for

¼ inch stock, but I haven't done it personally, so I can't tell you how well it'd work.

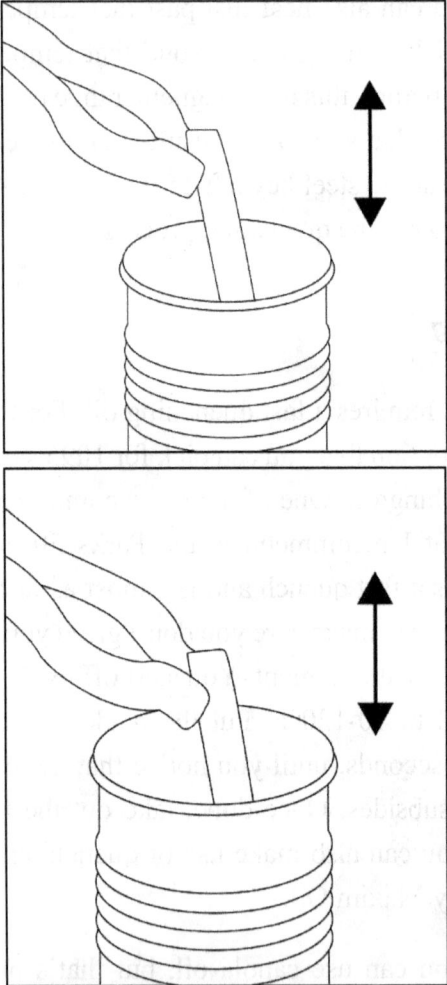

Figure 21 & 22: The dipping motion while quenching

When it comes to canola oil, preheat the oil to 135°F. Quench the knife in the liquid for about 10-15 seconds.

Tempering

The tempering process for 1095 steel is simple. Place the steel in an oven, and heat the steel to 500°F. We are trying to achieve about 59-60 HRC.

This will allow the knife to have good performance in most situations. However, I would still not recommend that you drop it. You might not destroy it entirely, but you might cause cracks to appear. Temper it twice for two hours each. Make sure that you are allowing it to cool. Bring it down to room temperature before you temper it again.

Cryo Treatment

Now, this step is not entirely necessary. However, it will improve the quality of the steel you are working with. If you like, you can skip this step altogether.

Soak the steel in temperatures ranging from -90°F to -290°F. The medium you should choose for cryo treatment should be liquid nitrogen. You need to ensure that you have introduced the metal to cryo treatment for about eight hours. For this, you can even soak the metal in liquid nitrogen overnight.

Eliminating Blemishes, Scaling, and Warps After Heat Treatment

To eliminate any of the blemishes that appear on the knife after the heat treatment process, you have to take the knife to the sander. At this point, it is especially important not to let your blade get hot on the grinder and ruin the heat treatment. Have a water drum near the grinder and dip the steel frequently. This way, if you feel that the knife is getting too hot, you can immediately dip it in the water to cool it.

But this time, there is a slight difference in how you approach the grinder. Usually, you hold the knife out horizontally and then grind away any of the materials still on the knife. This time, however, you are going to hold the knife vertically and then grind away.

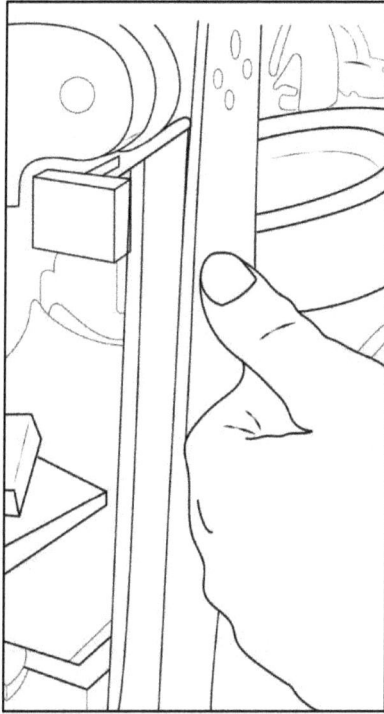

Figure 23: Vertical sanding is very effective in removing blemishes

One of the things that you will notice is that the sanding process might not remove some of the marks near the ricasso area.

For this, do not head back to the grinder again. If you can, take out the belt and use it manually to remove the blemishes and scaling. What you can do is use a long wooden stick. Tack one end of the belt onto one end of the wooden stick and the other end of the belt on the opposite end of the stick.

Bring the improvised belt to the knife and slowly sand away the remaining blemishes that you see.

Truing Your Knife

While most people assume that handheld blades are perfectly straight, the sad reality is that most blades are not straight. In fact, not only are most blades bent, but many are twisted as well.

So why are so many blades far from being straight? The answer lies in the fact that most people, knifemakers included, have not been taught to examine blades carefully.

Why is examining the blade important? This is because when you examine the blade, you get to check it for any defects or anomalies that you might have missed out earlier. Typically, when your blade is warped, then you might notice it easily. However, sometimes, you might have to check to make sure if there is a bend in your knife, especially when it might not be clear if there even exists a bend or not.

Here is how you do it.

Now we are going to use the idea of your dominant eye and non-dominant eye. Your dominant eye is the one you use to examine something. For example, if you want to peek at an object, then you often close one eye and open wide the other. The open eye is your dominant eye.

Let us start with the knife diagnosis.

To diagnose the straightness of a blade, hold the blade in such a way that the handle is furthest away from you. The edge of the knife should be towards the ground and the point aimed at your dominant eye. The non-dominant eye is closed. The back, or spine, of the knife should be in full view. Not only can the spine be examined for straightness this way, but straightness from the blade continuing to the end of the handle can be examined. Some blades are reasonably straight, only to slightly bend off somewhere along their bodies. Most will be bent one way or the other.

Sometimes, when you subject your knife to the heat treatment process, you might notice that it warps.

Warping happens for many reasons. Here are some of the causes for it:

- Heat treating of the blade is a delicate process. When you are subjecting it to all that heat, you might have to be careful how you distribute that heat. Sometimes, when bladesmiths use coal forge, they often forget to plunge the knife into the coal rather than simply place it on top of the distributed coals.

- During grinding, if cracks and splits are still present on the knife, then this can cause heated to be concentrated in certain areas, eventually causing the blade to warp.

- Make sure that you haven't skipped the tempering process or, for that matter, any process mentioned in the heat treatment of the blade. Remember that the only optional treatment is the cryo treatment. Everything else is important and should be followed in the steps mentioned above.

- Try and bring your metals as close to the non-magnetic limit as possible. This allows you to work with the metal better.

- Don't quench your knife in a sideways motion. Doing so will increase the chances of warping your knife.

Now let us assume that you have a warped knife. What can you do in this case? How can you straighten out the metal?

The process is fairly simple.

The first thing that you have to do is heat up the knife along the convex side of the curve. Once that is done, you can then hammer the knife or untwist it (or you can do both).

Let's look at each process.

Hammering

- When you are hammering, try to use the weight of your hammer to your advantage. If you find yourself holding on to your hammer with white knuckles, figure out how to get more comfortable.

- Find a rhythm and swing with your hammer instead of fighting against it. You can continue to hammer until you see the knife being straightened out. The index finger of the support hand is held at the exact spot to locate the problem. The blade is lowered to the anvil or your workbench, the convex side of the bend facing up, without removing the fingertip from the blade.

- A quick visual check confirms the exact spot on the blade where the hammer will strike. As the dominant hand reaches for the hammer, the support hand breaks contact with the blade to hold the blade by the handle in preparation for the coming hammer blow.

- A more forceful blow follows one very light strike to confirm the accuracy of the technique.

The blade is reexamined for results and repeated if necessary. In this technique, the emphasis is on proper diagnosis rather than on hammering technique.

Untwisting

- You can also untwist the knife. The best way to accomplish this is by using clamps to hold the knife properly while you use the tongs to straighten the knife. This process works to remove the warp easily, but it might require more strength from your end.

- Another way you can untwist the knife is by holding the knife in a vise. You have to make sure that the portion of the blade that has incurred the bend is placed in the center of the vise (since that is the part that is going to get twisted). Clamp down on the vise as tightly as possible, this will put pressure on the knife, and begin to straighten.

- If the twist is too small for your hammer, then you can make use of a metal rod. Make sure that the rod is thin enough to work with the twist that you have. Place the knife between a vise and clamp it hard. Once you have done that, position the rod so that it is aimed at the twist. Using your hammer, strike the rod softly until you can untwist the knife.

CHAPTER 7: FINGER GUARD
AND BOLSTER

Bolster

Making bolsters on full tang knives can seem daunting to the beginner. However, you can use the below process to make a bolster for your knife.

There are many materials that you can use to make bolsters. However, the recommended material for this purpose is brass. This allows you to make a bolster that is strong and sturdy and does the job well. Plus, it looks really good!

- When starting off, you will need to make markings for your bolster.

- Measuring allows you to focus your attention on the overall shape of the bolster and how you can work with it during the cutting phase. You can use a permanent marker to draw out the design on the brass piece. For the knife template that we have used, you can ideally go for a 1 inch to 1 1/2 inch bolster.

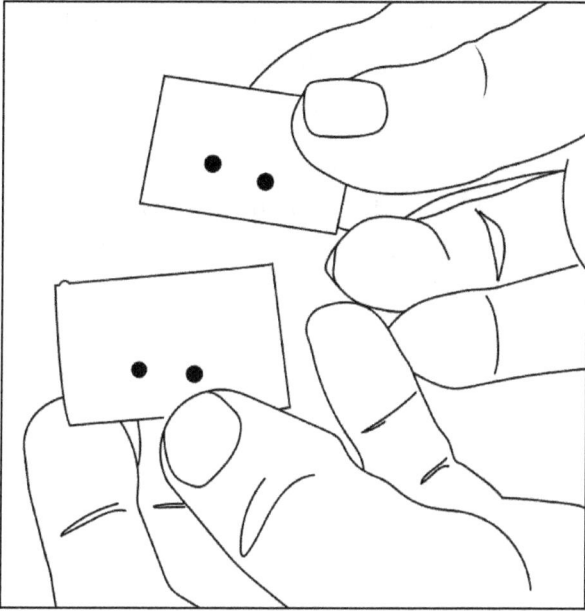

Figure 24: What your bolster pieces should look

- Then use your angle grinder or hacksaw and cut out the shape that you are aiming for.

- Now drill holes according to the template. If you have invested in a drill press, then you can make use of that as it provides you with more accuracy.

- When you are making a bolster, make sure that you don't use one single piece of brass. This will be difficult to add to the knife. Rather, find your design and create two pieces out of it. They can be clamped together on the knife. This makes it easier and you will be able to make adjustments easily if required.

- Once you have the 2-piece bolster ready, simply place one part of the bolster on the knife. Then place the other part on the other side of the knife.

- Once both pieces of the bolster have been lined up with the tang, it is time to use epoxy to glue them together.

- Smear the bolster pieces with epoxy and line up the pins with the holes in the tang.

- Push the bolster piece with the pins. Then hammer the pins through the tang and into the other bolster piece.

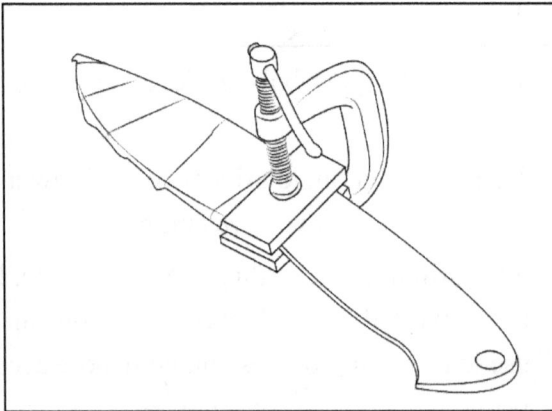

Figure 25: Clamped down bolster

- Now it's time to sand the bolster so that it flows perfectly with the tang.

- To sand it, simply use a 80-grit belt. Bring the bolster close to the belt sander and allow it to gently smoothen the part. Don't push the bolster too hard into the belt itself.

Finger Guard

There are numerous materials that you can use to make the finger guard. For our knife, we are going to use solid brass again. Make sure that the piece that you have with you is about 5 inches long and about 1 inch wide. If you can find a brass piece that is about ⅛ inches thick, then you have the perfect piece of metal to work with.

Time to get started with the process.

- Our first order of business is to mark the piece of metal with the outline of the guard. To do this, place the knife in such a way that the part where the blade of the knife begins falls on the piece of brass metal. This way, you have a part of the blade and a part of the tang on the brass piece of metal. Make sure that you place the knife as close to one end of the brass piece as possible. This allows you to use the other end of the brass as well. When you place the knife on one end of the brass piece, make sure you leave some space. About 1/2 inch of space is sufficient.

- This means that you have half an inch of space on one side of the knife. Measure a half-inch on the other side of the knife and mark that half-inch on the brass using a permanent marker. You now have a mark about 3-4 inches from one end of the brass piece (depending on the size of your knife).

- Now make two marks, one on either side of the tang. This is where the slot of the guard will go.

- Remove the knife.

- You will notice the lines you made for the tang won't extend across the knife. Take out your marker and extend both lines, so they cover the entire width of the knife.

- Once that is done, find the midpoint of these two tang lines. Draw a line that connects one midpoint to the other. Let us call this line the "midpoint line."

- Down use a drill press for the next step. You can also use your handheld drill, but you get more accuracy with a drill press. Clamp the piece of brass in a vise. Now, lower the drill press on one end of the midpoint line. Drill a hole.

- Move the piece of brass in such a way that the drill press extends the hold along the midpoint line. Eventually, the entire midpoint line will look like a gap that is almost the size of your tang.

- Once you are done, use a drill press to smooth out the gap as much as possible. When it is smooth, take it out and try and fit your tang through it. If you have the perfect fit, then your tang will go through the gap. If you don't, your tang might get stuck.

- If you would like to extend the hold slightly, do not use the drill press. Rather, use a narrow metal file to get the job done. Place the metal file on one end of the gap and run it back and forth a few times. Then do the same procedure for the other end of the gap. This way, you can extend the gap slightly. Remember that the guard should have a tight fit on the tang. It should not be loose, or the guard might come off.

- Once you have found out that the guard fits the knife snugly, you are then going to clean and polish the guard. To do this, you need to take a 600 grit sandpaper. Rub it on both sides of the guard. You need to rub it for a couple of minutes on one side and then the same duration on the other side.

- You can also use a grinder to complete the polish. You only need to subject the guard to the grinder for about 10 seconds or so to get the polish that you need.

- Once that is done, you can finish the guard by us- ing any regular brake cleaner. Spray some of the cleaner on the guard and then wipe it off using a cloth.

CHAPTER 8: HANDLE

To make the handle for your blade, you will attach two pieces of material to the outside of your tang. These pieces are called scales. You can make your scales out of a wide variety of natural and man-made materials. While there are some benefits and drawbacks of using certain materials, a lot of this comes down to personal preference. When choosing the material for a knife handle, you should take into consideration the environment and the kind of abuse your handle will need to take. If you're going to be hammering on your handle frequently during the process of batoning, it might not make sense to use a softwood that could be easily damaged. Changes in temperature and humidity will also make some natural materials shrink and swell, which could affect the integrity of your handle.

TIP: Using a bolster will shorten the length of handle scales that you will need for your handle. Take account of that when making the handle design.

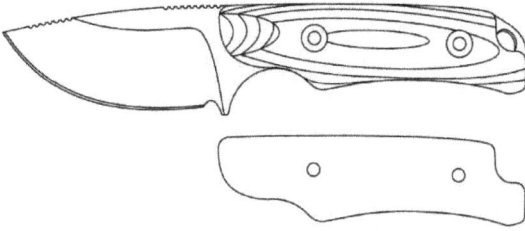

Figure 26: Handle scale template

Scale Material

Along with wood, one of the recommended knife handles you can use is made out of Micarta scales. Micarta is a form of synthetic material that is made out of certain kinds of fabrics, such as linen or canvas. They are usually soaked in resin. It is tough, lightweight, and makes a durable handle. When the handle is exposed to an oily or greasy liquid, however, it will make the Micarta a bit slippery. But that is a small drawback for an otherwise suitable material.

Another thing that you need to focus on is the pins that will be inserted into the tang and the handle. Pins are the pieces of thin, round metal that are inserted through holes to help hold the scales to a full-tang blade. These pins, once finished, will leave a small circle of metal visible on the handle. Pins can be made out of almost any kind of metal, depending on what you would like to see on your handle.

I would also recommend using Corby fasteners and Loveless bolts, if you have the taste for them.

Handle Making

- First step should be to shape the handle using the belt sander. In order to do this, you first choose the right material for the handle. You can either choose a piece of exotic wood with exquisite burls or use Micarta.

- You will now have a block of wood with you. The next step is to get the right dimensions for the handle.

- To get the right shape, you need to first understand that you actually need two pieces of blocks. These pieces will act as a clamp for your knife. To get the two pieces, you should use a hand saw to divide the block into two halves. You do this by running the saw straight down the middle of the handle. Try and make sure that the piece of material in front of you is evenly split in the middle. If you have uneven dimensions, then you might have trouble getting both halves of the material to form the right shape that you want.

- Each of the two halves should then be cut into the length of the handle. The handle length depends on the knife that you are making and the tang itself. But thankfully, we were prepared as we created a diagram to act as a blueprint for our knife. As we had seen earlier, the handle length that we are going to go with is 4 inches. However, try and cut it down to about 4 1/2 inches so that you can leave a little room for error.

- Once you have the correct lengths, you can then work on shaping the handle. If you have a rough idea for the handle, you can draw on the wood using a marker.

When you have the material for the handle (the scales) ready, you have to follow the below steps to complete making your own handle.

- Check to make sure your scales are perfectly flat. Figure out how you want your scales to sit on the blade.

- Make a mark on your scales to designate which side will be the inside. This will make it easier to keep track of which side to epoxy later on.

- Position one of your scales as it will sit on the blade. Clamp the blade and the scale together.

- Holding them secure, drill through the hole in your tang all the way through the scale.

- Push one of your pins through the holes, securing the two pieces. This will keep the existing holes lined up as you drill your second hole. Drill through the second hole in your tang, into the handle scale.

- If you have more than two pins in your handle, repeat this process, securing each new hole with a pin. Using your marker, trace the outline of your tang on the inside of the scale.

- Now remove all the pins and the scale that you just used.

- Use your hacksaw and follow the outline you made to cut out the rough shape of your handle. Put the scales back on the knife. It's okay if the scales are a little bigger than the tang. All you have to do to match them is sand down the profile of the handle to match the profile of the tang.

- Using coarse grit sandpaper, scratch the inside of your scales as well as the outside of your tang. Then, use acetone and a rag to wipe down all the surfaces of the scales. Next, we are going to mix the epoxy. One of the recommended epoxies that you can use is JB Weld. The ratio of epoxy to use will be 2:1. Which means you will mix two parts resin with one part hardener.

- Tape off the entire blade area of the knife, with some masking tape. You do not want to get epoxy on your blade.

- Using a clean and smooth stick, spread the epoxy all over the inside of each scale, and the tang. At this point, make sure you have enough epoxy that when you clamp the scales to the knife, the inside surface area is completely covered.

- Apply a little bit of epoxy on the ends of the pins. Put the pins through the first scale and into the tang. Fit the second scale from the other side. Make sure the pins go through the scale. Gently tap the pins with a hammer to make sure they don't get stuck and end up in the right position.

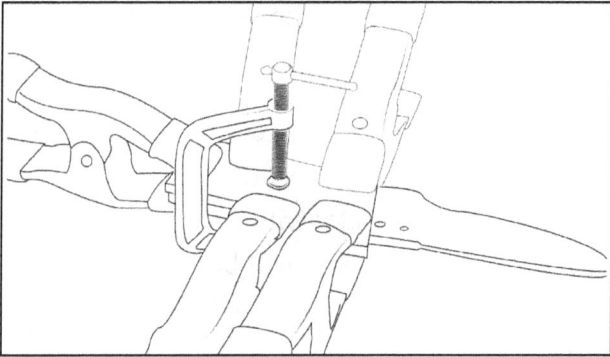

Figure 27: Clamp down everything nice and tight

- Clamp down on the handle, squeezing everything together securely. Wipe off any excess epoxy that runs out. Let your knife sit clamped overnight to dry.

- Once your epoxy has set, bring your knife back to the belt grinder. Clean up the edges and start shaping your handle.

- Test your grip as you go, take off more material wherever it puts pressure on your hand in an uncomfortable way.

- Once you're happy with your grip, take the belts down to a finer grit sandpaper. Finish up your handle by hand sanding with progressively finer grits until you get the finish you want.

Sanding and Shaping the Handle

Now that you have created the handle, you now have to get it shaped.

- Using the design on the wood as a guide, take your knife to the sander. For this purpose, you should use a sander with a belt that has 120-grit. With a 120-grit, you will be able to get the right shape and even add a smooth layer on your handle.

- Gently bring the handle to the sander. Work on the handle as you chip away the parts that lie outside the marked area. If you feel like you have to stop and examine the handle, do so.

- Continue working on the handle until you start seeing the shape that you had originally wanted to create. Work with the grit and then finish off the shape.

- Once you have done that, you might notice that the handle does not look 'finished.' It might have a rather rough body and plenty of wood shavings (if you are using wood) sticking out. At this point, you have to focus on giving a nice finish to the two pieces of the handle. Switch to 60 grit belt and then make your handle as smooth as much as possible.

- The next step into the sanding process is to make sure that you have sandpaper with the right grit.

- Before we even start the sanding process, you are going to have to protect your blade. Use a piece of leather to cover the blade. Make sure that the leather is soft inside so that it does not leave any marks on the blade. You can also make use of cloth, but the leather is tougher and resistant to sandpaper. If the cloth gets in the way of the sandpaper, then it might get messy with bits of stray cloth sticking to the sandpaper or revealing parts of the blade.

- Another reason for using a cover for the blade is to protect it from the vise. The clamps are going to hold the blade with the handle sticking out freely. In order to prevent any marks on the blade from the clamps, it is much better to wrap it using a cover (once again, preferably using leather).

- Now go ahead and clamp the blade and the handle ready for the sanding process.

- Start with an 80 grit sandpaper. When you are sanding the handle, you are going to move in a way that allows you to curve around it. This process allows you to cover the entire surface area of the handle.

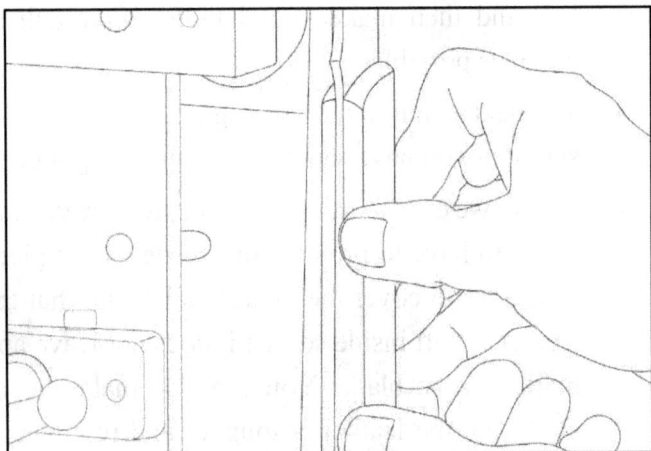

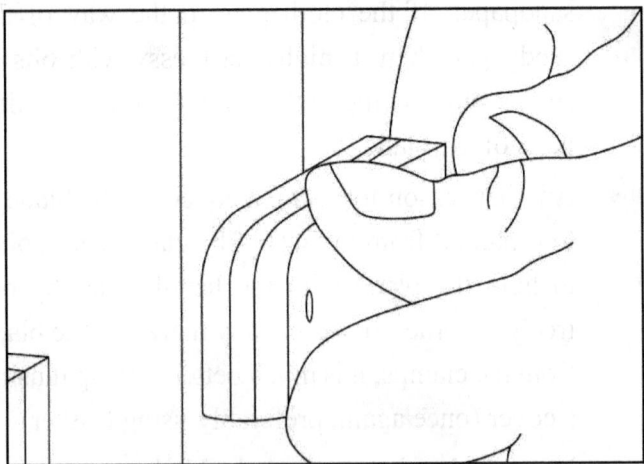

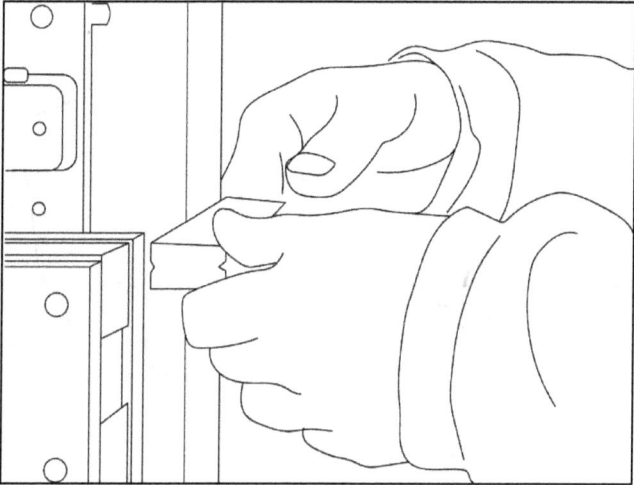

Figures 28 to 30: Proper sanding technique

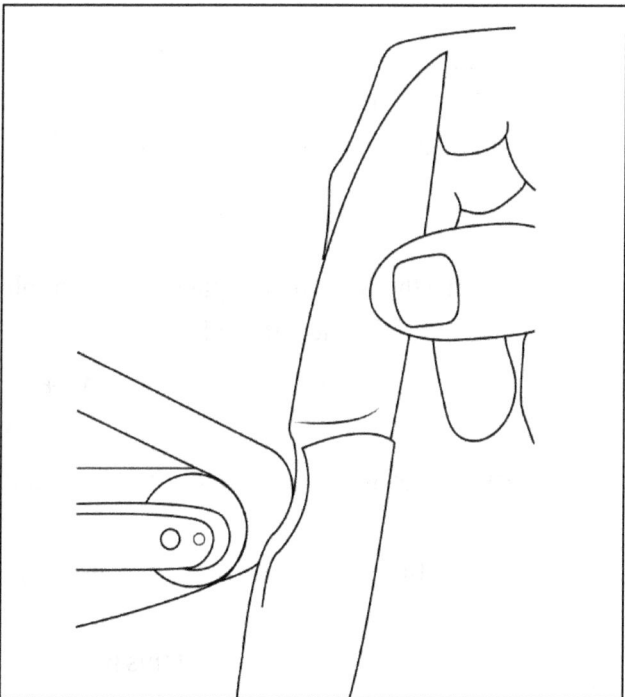

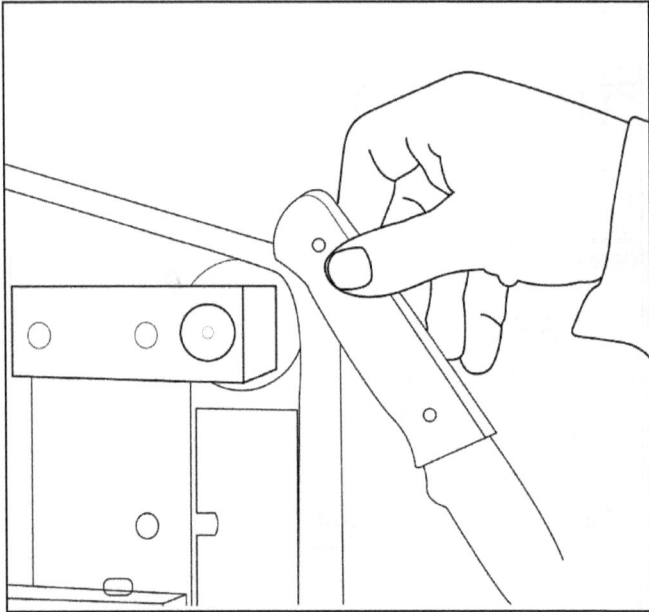

Figure 31 & 32: A small wheel attachment on a 2X72 grinder can be handy in shaping the curves of the handle

- Work with the 80 grit sandpaper for a couple of minutes on both sides of the handle.

- Once you have done that, shift the sandpaper to 240 grit.

- A few things to remember during the sanding process:

  o Take a little extra time when you are sanding the pins. If you don't spend enough time, they might 'dome' out a bit. What this means is that they bulge out of their holes because the area around them is getting sanded faster than them.

- o Be careful when you are sanding near the areas that have metal. If you sand those areas too much, then the metal will start protruding out, just like in the case with the pins.

- o After you have completed sanding using the 240 grit sandpaper, examine the handle and see if the results are according to your expectations. If you have to, redo the sanding process to get better results for your handle.

- Once the handle is fully ready, use Danish oil to polish and coat the entire handle.

CHAPTER 9: THE FINAL PROCESSES

Hand Satin Finishing for the Knife

- We had earlier worked on the handle, and this time, we are going to work on the blade. For finishing this knife, we are going to do a hand satin finish. We will basically replace deep scratches with finer ones, till the scratches are so fine that they aren't visible. If you want to sell your knife, then finishing it is necessary for your customer to feel good about his purchase.

- The best way to work with the knife is first clamping down a piece of board between a vise (make sure the board is narrow and more or less reaches the width of the knife blade). Then place the knife on top of the board and clamp the knife over there.

- Alternatively, cover the tang or handle of the knife with leather and then clamp the handle, leaving the blade projecting outward for you to work with your sandpaper.

- Use a little bit of WD40 and rub it along the blade. This elevates the sandpaper's cuts, and makes it last longer.

- Apply the WD40 on the belly of the knife. Whenever you are ready, take the sandpaper and place
  it on the blade that you are planning to finish.
  Then start moving the sandpaper along the length
  of the knife.

- You are going to sand the knife starting with an
  80 grit sandpaper. Then continue using progressively higher grits as you remove scratches from
  the coarser sand paper.

- Remember that when you are using sandpaper,
  you want to work at an angle. Imagine the knife
  is pointed away from you. You start with the tip
  of the knife and move the sandpaper side-to-side
  as you make your way up the knife's blade, towards yourself.

- Rather than directly sanding the paper from left
  to right, you can adjust the sandpaper to be at an
  angle. So when you move from side-to-side, it
  looks like the sandpaper is positioned at a
  roughly 45° angle. This allows you to cover more
  surface area while you are sanding.

- Sand both sides of the knife using the above process. Once you have finished sanding, then you
  can work on the bevels. When you start working
  on the bevels, start by covering them with a blue
  marker. This allows you to check if there are any
  spots on the bevel. In case of spots, you should

levelling should be done. Use a machined bar and dry paper to get the job done.

- Remember that the primary purpose of sanding is to remove any small scratch marks that might have appeared on the blade. In the end, when you have finished the sanding process, go over and check your work. Make sure that you are satisfied with the results. If you notice that there are still scratch marks, then take out the sandpaper and start working on it again.

- Don't be afraid to take time to remove the scratches. At this point, many knifemakers feel frustrated because they are so close to the end. They progress through the sanding process quickly to shoot through to the end. However, you should take your time. The fact that you are so close to finishing your knife might compel you to speed up the sanding process, but you should take your time with it.

- Using a progression of grits for sanding will fetch you better results than jumping grit sizes.

- When you have done everything right, you should be left with a hand satin finish.

Sharpening Your Knife

When you are working on the sharpening of your knife, you will come to realize that there are many sharpening tools out there in the market. Let us look at some of the tools that you can use for your sharpening process.

Understanding More About Knife Sharpening

The thinness of an edge makes it the most vulnerable part of your knife. This is also the part of the blade that takes the most beating. Every knife requires edge maintenance eventually, as even the best steel wears with time. The basic mechanics of sharpening remain the same, whether it's your blade's first edge or its hundredth.

Some grinds do best with a small, secondary bevel on the very edge. Other grinds, such as the Scandi grind, are sharpened by refining the original grind. This makes the Scandi grind a very easy grind to sharpen for beginner, as the angle needed is easy to determine.

Make sure you have good lighting before you start the sharpening process. As with grinding, your sharpening process involves starting with coarse grit and slowly moving down to finer and finer grits.

When sharpening, the key is to match the angle of the knife's edge to the sharpener. By keeping this angle consistent and moving your edge across finer and finer grits, you'll remove all the metal that won't make up the edge of your blade. The mechanics behind sharpening aren't difficult to understand, but good results take a skilled hand and refined attention to detail.

Knife Sharpening Technique

- Hold your knife flat on its side on your coarse stone. Lift the spine slightly so that the edge is resting on the stone at a sharp angle. Ideally, you should keep an angle of about 20-25°. However, you can choose the angle that works best for your knife.

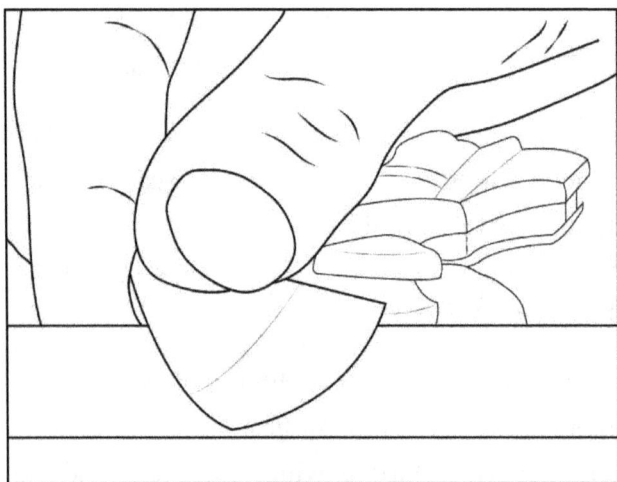

Figure 33: The correct sharpening angle

- Additionally, try and use a 1,000 grit stone in the beginning. This stone works well for beginners.

- Move the edge across the stone lightly, as if you were trying to slice off a very thin piece of it. Make sure the entire length of the blade, from the heel to the tip, comes in contact with the stone.

Always sharpen your blade in the direction away from you. Repeat this process several times, maintaining the same angle.

- As you remove steel from the edge, a tiny burr will eventually form on the opposite side of the edge. The burr is a rough, raised metal curl that results from grinding metal. The burr should appear evenly along the entire length of the edge. If you find that it is absent in an area on your edge, you aren't sharpening as much on that particular spot.

- Once you have a burr along the entire edge, switch sides. Repeat the process until you have a burr on the second side. If you have any chips in your blade, you'll have to continue grinding with a coarse stone until all the steel is removed past that chip. Once you have set your edge with the coarse stone, move to a finer grit stone and repeat the whole process.

- Use a 'strop' method to remove the final burr on the blade's edge. Stropping involves the same motion that is used while sharpening on stone, but in reverse. Instead of cutting forward, the blade is drawn back, dragging the edge. Use light pressure and make several passes, alternating sides. Stropping makes sure that the sharpness of your blade remains for a long time. Note that stropping stones are different from sharpening stones, and you should invest in one if you are serious about knifemaking.

TIP: Here is an important tip to remember while sharpening your knife. Make sure that you clean your knife before you bring it to the stone or strop block that you are using. If you don't, there are chances that you might contaminate the stone that you are working on more.

Testing the Edge

There are many ways to test your edge to see if it is sharp enough. Everyone has their own favorite method, from seeing how well the blade cuts through the paper to carefully drawing the edge along a fingernail. In my opinion, the best way to test the edge is to use it. The efficiency at which it completes any given task helps to determine whether or not the edge was sharpened properly.

Another method to test out the edge (especially for our hunting knife) is to run the knife through a paper towel. If you can drag the knife through the towel without much resistance and by its own, then you have yourself a well-made and sharp knife.

BONUS CHAPTER: MAKING TONGS

Now we are going to focus on making simple, but effective tongs that you can use in your metal works.

There are numerous materials that you can use for making your tongs. One of the more common materials that you can get your hands on is rebar. You can get one that is about 3 feet long and is about 1/2 thick.

- The first thing that you have to do is find the center of the bar. The make a mark on the bar using a permanent marker. From the center, measure 3 inches to the right and then 3 inches to the left. Call these two points A1 and A2.

- Now make an indent on A1 and A2. You can do this by placing the bar on the edge of your anvil and lightly tapping it. Using only the two indentations, you have a space of 6 inches in the center.

- Now take the rebar to your forge and heat the marked part.

- Bring over the bar back to the anvil and flatten the heated part. Strike on it using your hammer and make sure the entire 6 inches of the center is getting flatter. When they seem flat enough while still having about an inch or so of thickness, take the part back to the forge and heat it.

- Whenever you heat the bar, make sure that you are heating it to the yellow range temperature.

- Once the bar has been heated, bring it back to the anvil. Now we are going to use the horn part of the anvil (the little protrusion at the front). Place the heated part on the horn and then start striking the bar to bend it in such a manner, the two ends of the bar are going to meet. Strike the bar evenly so that when you create your tongs, you don't receive an uneven or awkward shape.

- Eventually, you are aiming to bring the two ends to a point where they look like they are making a 'U' shape. When you have achieved the shape, you have completed the first part of the process.

- Heat up one end of the bar and bring it to the yellow range temperature. Head over to the clamp and hold it firmly with the end pointing upwards. Now we are going to use a chisel and place it in the center of the base of the bar. Essentially, we are going to split the end. Line up the chisel in the center and strike it using a hammer. You can make the split as deep as possible but make sure that you do not overdo it.

- Repeat the above process on the other end of the bar as well. Basically heat it, clamp with the end pointing upwards and strike it so that you split it.

- The next step is going to involve the ends as well. This time, heat the ends and using the horn of the anvil, bend them inwards at a 90-degree angle.

- Start with one end of the bar and then work on the other end. In the end, you have both bars bending inwards. This formation becomes the hands of the tongs, holding any piece of material between them.

- Go back to the center of the bar. Heat it again and bring it back to the yellow temperature range. Remember those indentations that you had made? Choose a point slightly close to those indentations and strike the bar so that it starts bending inwards.

- Do the same with the other side of the bar as well.

- Now it looks like you have an indentation on both sides of the bar, close to where it bends to create the 'U' shape.

- When you have done that, you now have a very rudimentary form of tongs. You can use the bent ends to clamp down on any piece of metal and then bring it out of the forge easily.

- One tip for the process is to replace the rebar with coil springs because coil springs have that extra bit of elasticity to allow you to squeeze the two ends of the bar together.

- When you are using coil springs, make sure that you first start by cutting 3 feet of metal off the spring. Once done, you should then straighten the metal. Straightening a coil spring is relatively easy as the metal itself does not provide much resistance. Heat the parts that are bent and using a hammer, strike the metal lightly until it straightens.

- You can make use of many other metals for this, but rebar is easy to get while a coil spring makes for effective and flexible tongs.

CONCLUSION

Bladesmithing is a satisfying process. The hard work that you put into it reaps some incredible results. Of course, it all depends on the efforts and the time you put into it.

Do not worry about the mistakes that you make. Every mistake is a learning curve for you. It is for this reason that the metals you are using in this book are easy to work with. Even if you do not get the right shape or perform a mistake during the grinding process, you don't have to worry about the metal. You can either choose to correct it easily or get yourself the metal (as it is fairly easy to get).

There really is something special about a razor-sharp knife that, once experienced, will be hard to live without. The superior cutting performance also factors in blade shape and geometry and ease of sharpening.

Do go through this book carefully before you jump in on the heat treatment processes. Make sure you understand the concepts. Most importantly, be careful when you are working with metals.

Always put yourself first. Are you in a safe environment? Are you keeping yourself protected? Are you staying a safe distance from fire and other harmful objects?

Remember, there is no point in trying something when you are not feeling safe.

Another factor that you must consider is that bladesmithing is a fairly time-consuming process. When you are aware of this, you might decide how best to approach the various processes. That is why, make sure that you are comfortable with one process before you carry on to the next. For example, if you feel like you are not doing the quenching right, then refer to this book and try again. Try not to skip to another step if you haven't properly performed the previous step.

You might find yourself physically exerting force on the metals you are working with. Sometimes, this might become a little uncomfortable. Rest your hands if you think that you are putting too much strain than is absolutely necessary. However, remember that if you follow the instructions in this book, then you won't have to strike the metal too hard.

Simply watching your blade come to life is one of the most enjoyable sensations you can experience. I am sincerely hoping that you have such an exquisite feeling yourself.

Keep yourself protected and enjoy a wonderful bladesmithing process.

And if you feel like it, drop a quick review for this book as well. I would really appreciate it.

REFERENCES

Blandford, P. (2006). *Blacksmithing projects*. Mineola, N.Y.: Dover Publications.

Blandford, P. (2010). Practical blacksmithing and Metalworking. New York: TAB.

Parkinson, P. (2001). *The Artist Blacksmith*. Crowood Press.

Streeter, D. (2008). *Professional Smithing*. Lakeville, Minnesota: Astragal Press.

www.ingramcontent.com/pod-product-compliance
Lightning Source LLC
Chambersburg PA
CBHW071704210326
41597CB00017B/2323